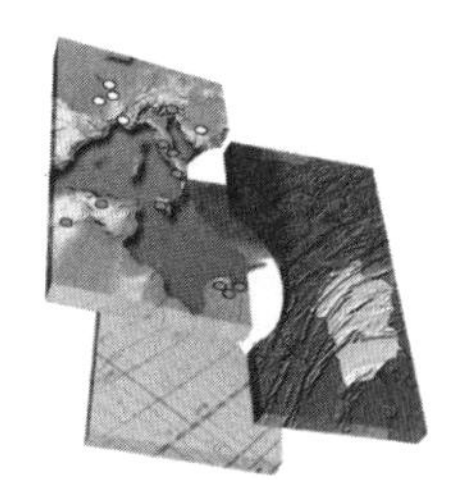

Developing GIS Solutions with MapObjects™ and Visual Basic®

Bruce A. Ralston

ONWORD PRESS
THOMSON LEARNING

Australia Canada Mexico Singapore United Kingdom United States

Developing GIS Solutions with MapObjects™ and Visual Basic®

By Bruce A. Ralston

Publisher:
Alar Elken

Executive Editor:
Sandy Clark

Acquisitions Editor:
James Gish

Managing Editor:
Carol Leyba

Development Editor:
Daril Bentley

Editorial Assistant:
Jaimie Wetzel

Executive Marketing Manager:
Maura Theriault

Executive Production Manager:
Mary Ellen Black

Production Manager:
Larry Main

Manufacturing Coordinator:
Betsy Hough

Technology Project Manager:
David Porush

Cover Design:
Cammi Noah

SAN 694-0269

Printed in Canada
10 9 8 7 6 5 4 3 2 1

For more information, contact:
OnWord Press
An imprint of Thomson Learning
Box 15-015
Albany, New York 12212-15015

Or find us on the World Wide Web at
http://www.onwordpress.com

Library of Congress Cataloging-in-Publication Data
is available for this title
ISBN: 0-7668-5438-8

About the Author

Dr. Bruce Ralston is a professor in and head of the Department of Geography at the University of Tennessee, Knoxville. He holds multiple awards, including the Edward L. Ullman Award for Outstanding Contributions to the Field of Transportation Geography, and the Applied Geography Award, both from the Association of American Geographers. He is also a consultant to various public and private sector organizations and President of GIS Tools, Inc.

Acknowledgments

I want to thank the many students who had the desire to learn this material. One of the luckiest things that can happen to a professor is to have students whose enthusiasm for learning pushes him to also learn. I have been extremely fortunate at the University of Tennessee to have a steady stream of such students. They have helped me learn many new areas, including the topics covered in this book. Grateful acknowledgment is also made to the team at OnWord Press.

For James and Catherine

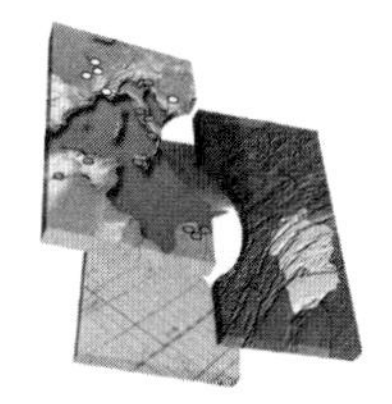

Contents

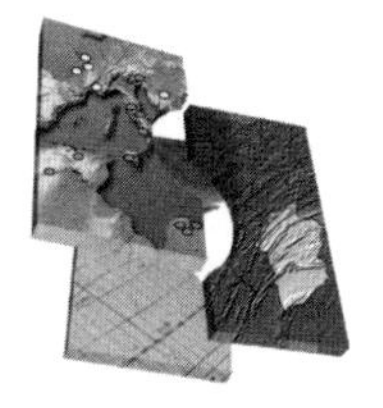

Introduction

If you have ever tried to learn a foreign language, you know how helpful a dictionary can be. You also know that it is virtually impossible to master a language simply by studying a dictionary or taking classes. Ultimately, you have to immerse yourself in the culture, finding yourself in situations in which you have to use the language to be successful, such as ordering a meal or obtaining directions. A dictionary can help, but it is not enough.

■■ Purpose and Approach

The aforementioned is the problem with help files and, quite frankly, a lot of programming books. They are more like dictionaries. They help you look up specific things once you are conversant in the language, but they do not put you into the culture. Even the example programs in the MapObjects Help file, although useful, provide only "snippets of conversations," not a complete discourse in the original language.

This book attempts to "get you into" the Visual Basic/MapObjects culture. The book begins by examining some simple programs, and by Chapter 4 you will be building a program—a conversation with a user—that becomes more complex in subsequent chapters. In most chapters you will learn a new aspect of Visual Basic programming and then use that knowledge to exploit the potential of MapObjects.

■■ Audience and Prerequisites

This book was developed for users of GIS who want to learn how to develop their own applications using Visual Basic and MapObjects. Although basic aspects of Visual Basic programming are presented in the text, this is not meant to be a comprehensive introduction to computer programming. The goal of the book is to help readers learn enough about VB and MO so that they can develop their own programs and implement them as either stand-alone applications or as the basis for web-based GIS. Experience with other ESRI products, such as Arc or ArcView, is helpful.

There are some software prerequisites for using this book. Readers should have access to Visual Basic 6, MapObjects 2.0 or later, and (if you wish to implement the web-based application described in Chapter 15) MapObjects IMS. Readers who do not have these ESRI products can download 90-day evaluation copies at *www.esri.com/software/mapobjects/download.html* and *www.esri.com/software/mapobjects/ims/eval.html.* Chapter 14 requires the user to have a web server application that supports PERL scripts.

How To Use This Book

The book contains a series of example programs that build from a very simple map display application in Chapter 4 to advanced applications in later chapters. Starting with Chapter 4, each succeeding chapter will add new functionality to the program developed in the previous chapters. Each program developed in the text can be found on the companion CD-ROM, so that readers can either enter the code that is described in the text or load the corresponding sample programs from the companion CD-ROM.

Content and Structure

This book can be broken into four parts. Chapters 1 through 3 provide a brief introduction to programming with Active X components and Visual Basic (VB). Readers familiar with VB may wish to skim these chapters. The remaining chapters deal with developing GIS programs. In each chapter, a new aspect of VB is discussed, along with new GIS properties that can be found in MapObjects (MO). The VB discussed in the chapter is then combined with MO objects to add more GIS functionality to the program developed in the chapters. Chapters 4 through 7 deal with managing map layers.

Chapter 7 deals with identifying elements of map layers and working with selected sets. Chapters 8 through 11 cover thematic mapping. Chapter 12 deals with collections, classes, and theme-on-theme selections. Chapters 13 through 15 cover web-based GIS. Chapter 13 provides a brief introduction to HTML, whereas Chapters 14 and 15 present two approaches to serving maps on the Web. The final three chapters (16 through 18) deal with map projections, coordinate systems, and buffering and overlay.

Book Features and Conventions

This edition includes a companion CD-ROM at the back of the book (see "About the Companion CD-ROM" at the end of this introduction). Throughout the book you will see references to the companion CD-ROM. It is there that you will find VB projects, data sets, DLLs, and other utilities discussed in the text.

Italic font in regular text is used to distinguish certain command names, code elements, file names, directory and path names, user input, and similar items. Italic is also used to highlight terms and for emphasis.

The following is an example of the monospaced font used for code examples (i.e., command statements) and computer/operating system responses, as well as passages of programming script.

```
var myimage = InternetExplorer ? parent.
cell : parent.document.embeds[0];
```

The following are the design conventions used for various "working parts" of the text. In addition to these, you will find that the text incorporates many exercises and examples.

NOTE: *Information on features and tasks that requires emphasis or that is not immediately obvious appears in notes.*

TIP: *Tips on command usage, shortcuts, and other information aimed at saving you time and work appear like this.*

CD-ROM NOTE: *These notes point to files and directories on the companion CD-ROM that supplement the text via visual examples and further information on a particular topic.*

About the Companion CD-ROM

The companion CD-ROM consists of four directories, each of which contains material used in the text. Each directory and its content are described in the following.

- *MO2*: This directory contains all VB projects referenced in the text. Whenever you see a CD-ROM Note that references a particular directory, it will be found under this directory. These projects can be opened and executed in Visual Basic.
- *Utility*: This directory contains a DLL for converting bitmap files to *jpg* files. It is used in Chapter 14 to develop web-based GIS without a commercial IMS product.
- *Web*: This directory contains materials used for web-based GIS. These include HTML pages, two PERL scripts, Active Server Pages (ASP) scripts, and a *jpg* file of Tennessee. The use of these files is discussed in Chapters 13 through 15.
- *Shapes*: The shapes used in the examples are stored here. There are two directories under the *Shapes* directory. *USA* contains shapes for the United States, and *Knox* contains shapes for Knox County, Tennessee. The shapes in the *USA* subdirectory are used in Chapters 4 through 15, and the shapes in the *Knox* subdirectory are used in chapters 16 through 18.

Chapter 1

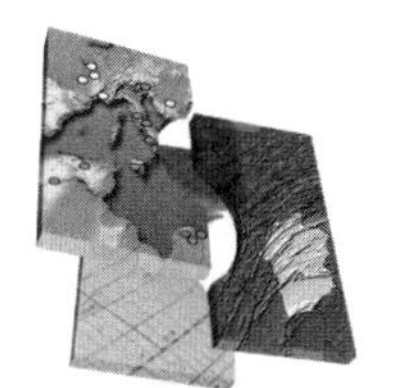

Entering the VB/MO Culture

■■ Introduction

Developing GIS solutions requires writing code that can manage and manipulate geographic data. Advances in software development tools have accompanied the rapid acceptance of GIS. Two advances in software development have been the development of object-oriented programming (OOP) languages and the use of components for the distribution of objects that can be used with OOP languages. Visual Basic (VB) is a language that allows users to develop and manipulate objects. MapObjects contains a set of objects of interest to GIS application developers. Together, these tools make it possible to develop OOP-based GIS implementations.

■■ An Introduction to OOP Implementations

To understand OOP you must understand objects. Objects have *properties* and *methods*. It is important that you understand the difference between these. First, however, you need to understand the concepts "class" and "object." A *class* is an abstract (meaning non-discrete) term, such as *human being*. An *object* is a discrete realization or instance of a class (i.e., by analogy, a particular human being). It is in the class that *properties* and *methods* are defined. Every object (and therefore its properties and methods) is the realization of a class, just as every person is the realization of the concept "human being."

Toward understanding the difference between properties and methods, let's use another real-world analogy: a bicycle. A bicycle (an "object") has many properties, among them color, size, weight, number of tires, brand, and type of spokes. Applying this analogy to scripting, if you wanted to determine a property (such as color or brand) of the object (bicycle), you would typically use a *Get* command, as in

```
color = bicycle.GetColor()
```

or

```
brand = bicycle.GetBrand()
```

The first instance returns a color, which might be represented by a string (such as BLUE) or a number (such as 0xFF000, which is the hex number—a number in base 16—for orange). The second instance is more likely to return information represented as a string, such as TREK (a brand of bicycle). The point is: *You must know what type of variable your* Get *request is returning.* At this point you might want to try adjunct exercise 1-1, which follows.

Adjunct Exercise 1-1: Determining Variables Returned by the Get Request

NOTE*: This exercise assumes you have ArcView 3.*

To determine variables returned by the *Get* request, perform the following steps.

1. Start ArcView.
2. Go to the help file, click on the INDEX tab, and type *GET.*
3. Scroll down and note that there are a lot of entries. Click on the FIND tab, and type *get* (lowercase, as this is case sensitive).

How many "get" topics have been found? Not all of these will be for *Get* requests, but, as you saw from the INDEX tab, a lot of them will be.

You now know that you can find an object's properties (like a map's extent or projection, or a layer's default symbol) with a *Get* request. Curiously, however, not all *Get* requests use the word *Get*!

Returning to our bicycle example, suppose you were ordering a new bike. In this case, you might want to order a specific color. The sister command of *Get* is *Set*, as in

```
bicycle.SetColor(BLUE)
```

Here, the color of the bicycle has been set to *BLUE*. Note that there is no equals sign (or more accurately "assignment operator") in this statement. (At this point you might want to try adjunct exercise 1-2, which follows.)

Adjunct Exercise 1-2: Determining Variables Returned by the Set Request

NOTE: *This exercise assumes you have ArcView 3.*

To determine variables returned by the *Set* request, perform the following steps.

1. Start ArcView.
2. Go to the help file, click on the INDEX tab, and type *SET*.
3. Scroll down and note that there are a lot of entries. Click on the FIND tab, and type *set* (lowercase, as this is case sensitive).

How many "set" topics have been found? Not all of these will be for *Set* requests, but, as you saw from the INDEX tab, a lot of them will be.

One last thing about *Set*: so far you have been looking at OOP languages in the abstract. Later in the book, you will see in regard to VB that there are two related ways of setting values: *Set* and *Let*. You do not need to worry about the difference between these just yet.

Some properties cannot be changed. For example, in regard to the bicycle example, you might not allow the following:

```
bicycle.SetNumTires(3)
```

Bicycles, by definition, must have two (and not more or less than two) tires. If there are three tires, it is a tricycle, not a bicycle! (One tire, and it is a unicycle.) This points out a very important rule, expressed in the following Note.

***NOTE**: With some properties you have read* (Get) *and write* (Set) *access. With others, you only have read* (Get) *access. Still others only have write* (Set) *access.*

Finally, consider the tires themselves. They, too, are objects. They, too, have properties and methods. When you want to inflate the tires, the object of interest is the tire, not the bicycle. That is,

```
bicyle.SetPressure(90)
```

may be illegal, but

```
tires.SetPressure(90)
```

may be fine.

This points out yet another important lesson, expressed in the following Note.

***NOTE**: You must know which properties and methods go with which object.*

This is often confusing and seemingly counterintuitive, but it is extremely important. Consider another example. Suppose you had an object named *HOUSE.* The house object contains several other objects, one of which might be *BATHTUB.* A command such as

```
bathtub.SetWaterLevel(top)
```

might be fine, but

```
house.SetWaterLevel(top)
```

might be disastrous! However,

```
bathtub.SetCleanToday(TRUE)
```

and

```
house.SetCleanToday(TRUE)
```

might both be legal and desirable.

In summary, you know that there are things called objects, which have properties and that can consist of other objects. You can determine the value of those properties with *Get* commands, and may be able to change those properties with *Set* commands.

Let's turn to methods. Methods are the things objects do. For example, a "person" object might have "eat," "sleep," "move," and "rest" methods. Our "bicycle" object might have "move" and "stop" methods. A map object can be drawn or scaled. A layer object can be turned off or on. In fact, you have already encountered methods. For example, to redraw a map in VB/MO, you would issue the command *themap.Refresh.*

Working in Visual Basic

Let's take a look at VB. As stated, VB uses an integrated development environment (IDE) that facilitates a "trial-and-error" approach to program development. That is, VB programs are developed in the IDE and can be tested without compiling the program or leaving the IDE. To start the IDE, you need only start VB (see figure 1-1).

CD-ROM NOTE: *The VB project in the* Chapter 1_1 *directory on the companion CD-ROM starts here.*

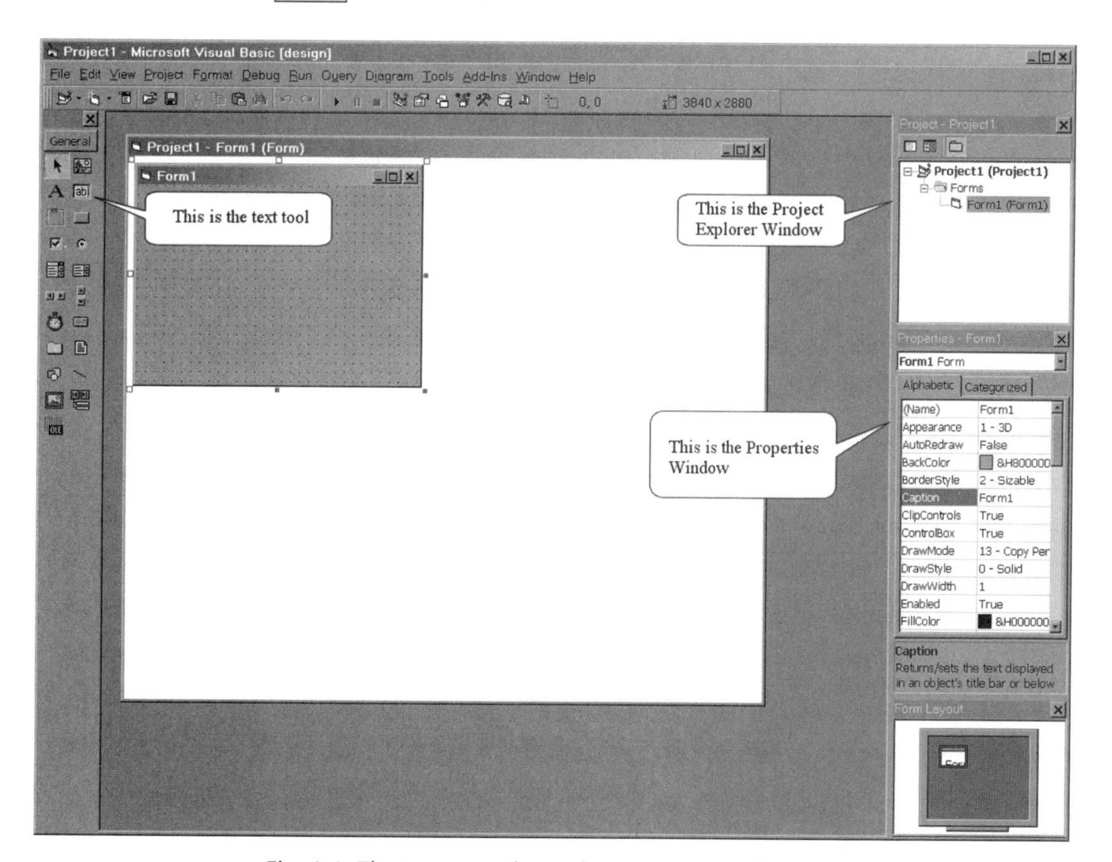

Fig. 1-1. The Integrated Development Environment (IDE).

The environment consists of a menu bar, context menus, toolbars, toolbox, the Project Explorer window, the Properties window, the Object Browser, the Form Designer, the Code Error window, the Form Layout windows, and the Watch and debugging windows. These interface elements are described in material to follow. Your screen may not look like that shown in figure 1-1. Some other windows may be present, or some that are labeled in the illustration may be missing. Do not be concerned, as this depends on how VB was last configured.

Forms are the dialog windows on which you place controls. Associated with each form is a code page—the page where your program resides. You can place controls (e.g., command buttons, text boxes, and radio buttons) on the form. The code corresponding to the controls usually resides in the code page for the form on which the control is placed. You can change from the multiple-document interface (shown in figure 1-1) to a single-document interface. You can also change window positions, menu docking, and the like. It is up to you to find out which environment settings you like best. The multiple-document interface, as shown in figure 1-1, is used throughout this text. Let's try a simple VB program.

1. Select the Text Box tool and draw a text box on the form (see figure 1-2).

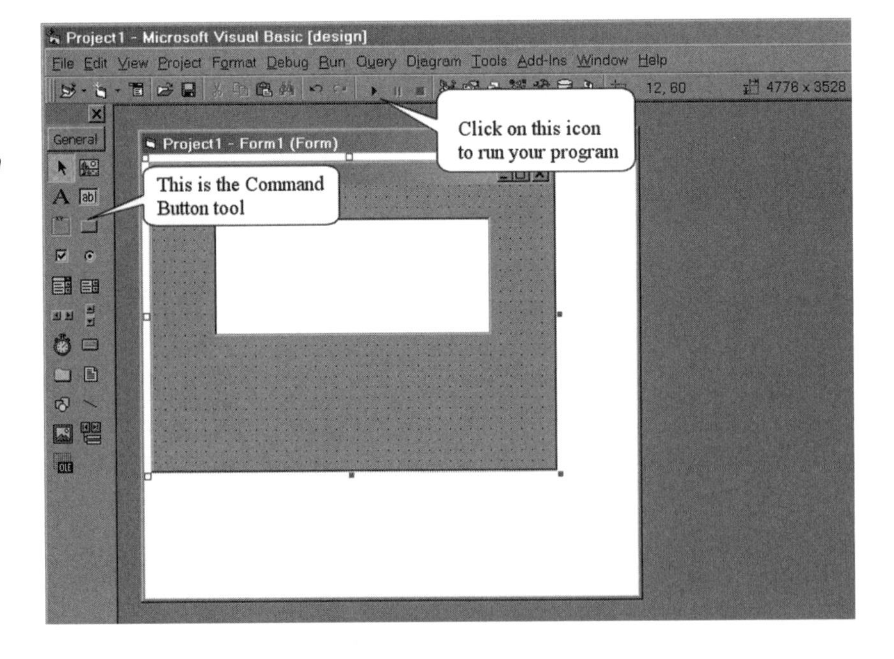

Fig. 1-2. Click on the Run icon.

2 Run the program. From the menu bar, click on Run and then on Start. Alternatively, click on the Run icon (see figure 1-2).

3 You will be asked to save your form and project. Do so.

The program will then start. It seems quite easy. (Okay, so the program does not do anything, but it still works.)

4 To stop the program, click on the Close Window icon (the X in the upper right-hand corner). Alternatively, in the IDE, click on the Stop Program icon (two icons to the right of the Run icon).

Now let's add a command button. Your IDE should now look like that shown in figure 1-3.

5 Go to the Properties window. (If you cannot find the Properties window, click on View, and then on Properties. Alternatively, press the F4 button on your keyboard.)

6 Once you have located the Properties window, select the form from the drop-down list. This will list the form properties. Find the Caption property.

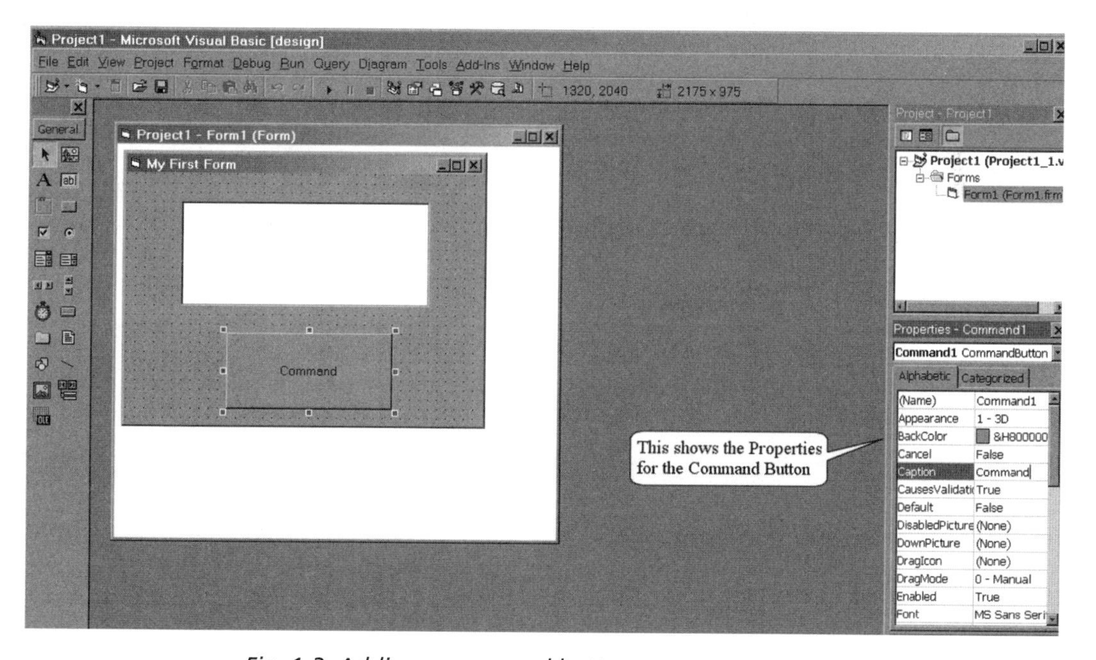

Fig. 1-3. Adding a command button.

Let's change the form caption.

7 Click on the current caption (*Form1*) and change it to *My First Form.*

8 Select the Command button from the drop-down list in the Properties window (or click on the Command button on your form). In its Properties window, change the caption to *Push Me—I Dare Ya!.*

Note that you are setting properties of objects. You can do this in the Properties window or in a program. In fact, the captions may change, depending on the events. All you can set here are the initial values of these properties.

9 In the Text Box properties, set the Text to empty, and then double click on the form (but not on the Command button or text box).

What happens? Your screen should look like that shown in figure 1-4.

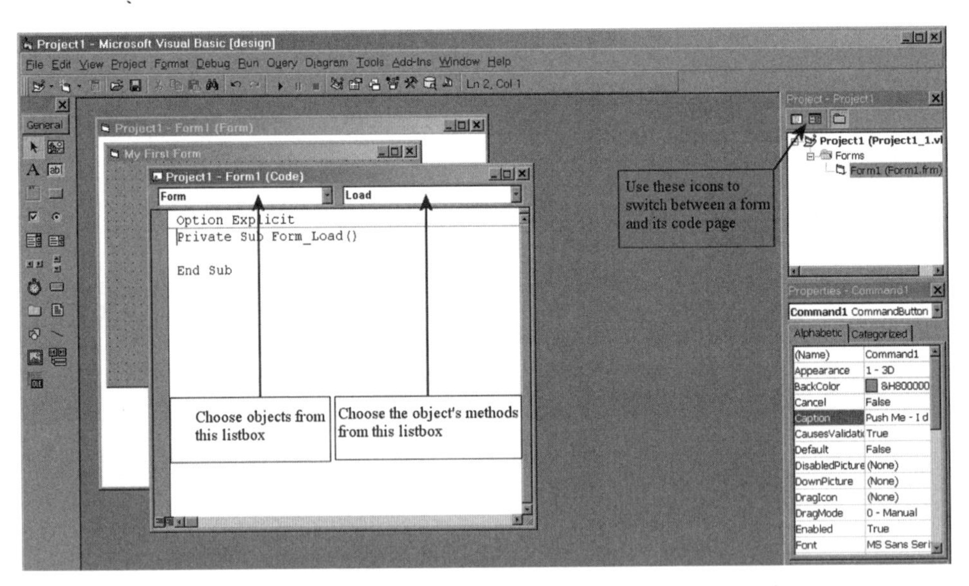

Fig. 1-4. Double clicking on the form brings up a code window.

10 Go back to the form and double click on the Command button.

What happens? The form's code window appears and a sub (short for subroutine) named *Command1_Click* appears. It is here you type the code that determines what happens when the user clicks on the Command button.

11 In the sub titled *Command1_Click*, type in the following:

```
Form1.
```

What happens when you type the period? Your screen should look like that shown in figure 1-5.

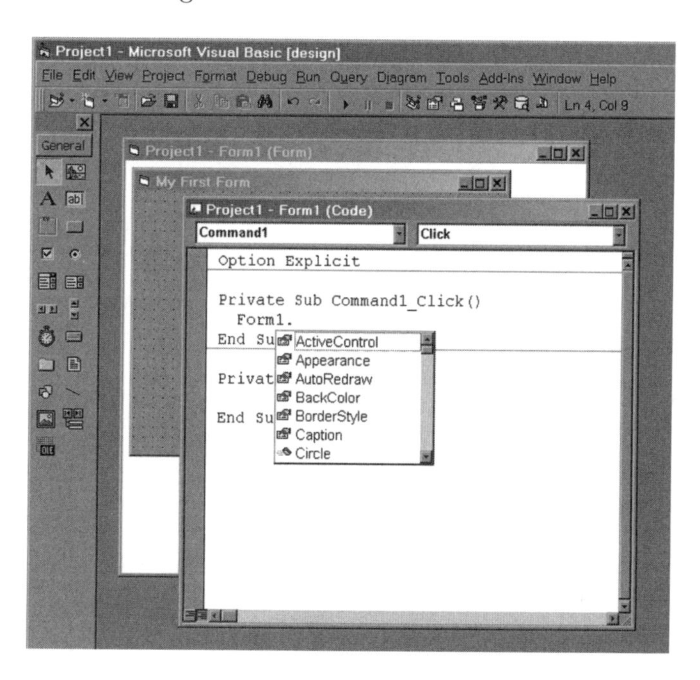

Fig. 1-5. The context-sensitive object properties and methods list.

All properties you can set/get and methods you can execute are listed.

12 Enter the following text.

```
Private Sub Command1_Click()
  Form1.Caption = "Hey--You hit me!"
  Text1.Text = "Watch it with that mouse"
  Command1.Caption = "Don't do that again"
End Sub
```

13 Run the program.

What happens when you click on the Command button? The result is shown in figure 1-6.

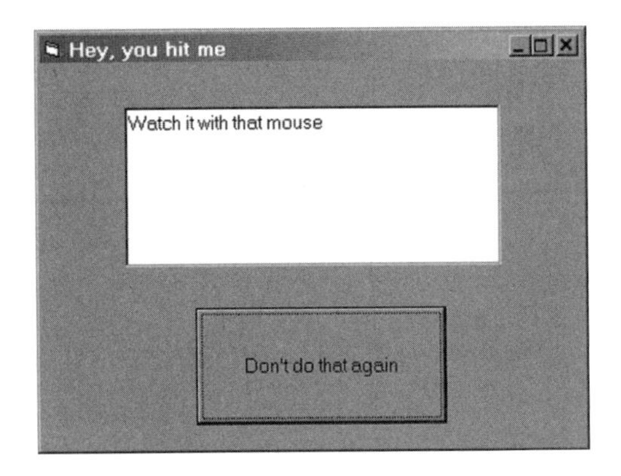

Fig. 1-6. Result of running the program.

14 Close your program.

The program currently consists of one form. The form has an object window (where you draw in the controls) and a code window (where you type in code). You can switch back and forth between these windows using the two leftmost icons in the Project Explorer window.

The icons at the top allow you to see the current object's code window, the current object's object window (the form itself), and all forms and other items (described later in the book) that constitute the project. Currently, the project only has one form in it. Figure 1-7 shows the current project's Project Explorer window.

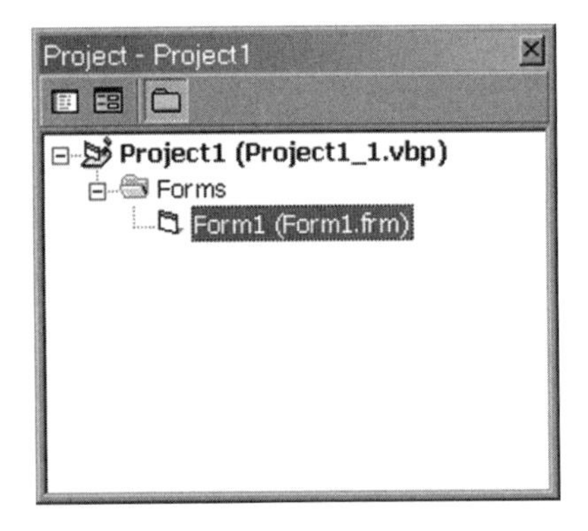

Fig. 1-7. The Project Explorer window.

CD-ROM NOTE: *The VB project in the* Chapter 1_2 *directory on the companion CD-ROM starts here.*

Now let's add a second form. The second form will pop up when the user presses the Command button on form *Form1*. The user will type a line in the text box on form *Form2*, and then click on an OK button. The line typed in form *Form2*'s text box will appear in form *Form1*'s text box, but in all uppercase letters.

To add a second form, perform the following steps.

15 Click on Project, and then on Add Form.

A dialog will appear allowing you to select the type of form you wish to add.

16 Select Form, and add a text box and a command button to this form.

The Command button should have the caption *OK*. The text box should initially be empty. You can set these in the Properties window for each of these controls. Your screen should look like that shown in figure 1-8.

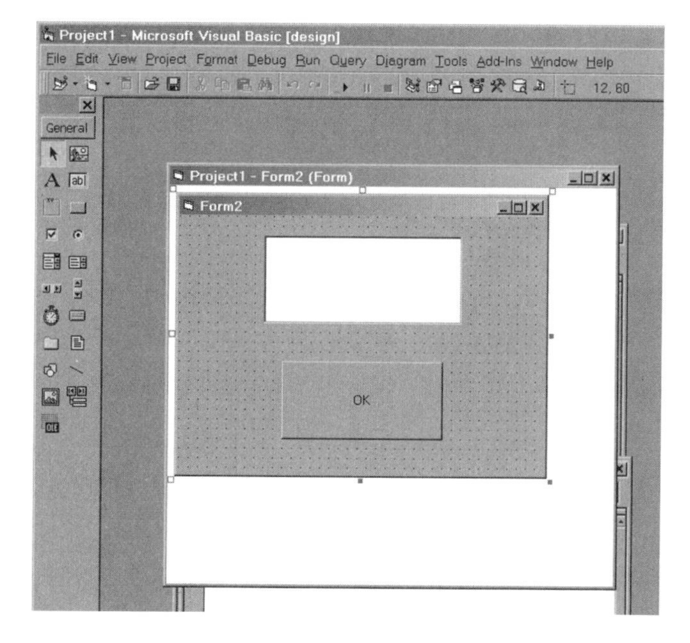

Fig. 1-8. The Form2 *layout.*

This is a good time to consider an important aspect of dialogs. When the user clicks on the button on the first form, the second form appears. The question is whether the second form should be *modal*. A form is modal if the user cannot access another form in the program without first closing the current form. In this example, if form *Form2* is modal, the user cannot return to form *Form1* without first closing form *Form2*.

17 Go to form *Form1*'s code page and change the *Command1_Click* code to the following (make sure you are in form *Form1*'s code window).

```
Private Sub Command1_Click()
 Form2.Show vbModal
End Sub
```

18 Run the program.

19 When the second form comes up, click on the first form.

What happens? (At this point, to stop the program you will have to click on the Close Window icon in each form.)

20 Change the *Command1_Click* event code to the following.

```
Private Sub Command1_Click()
 Form2.Show
End Sub
```

21 Run the program again.

This time you can switch back and forth between windows. Some forms you will want as modal, others as modeless. If you have worked with Windows-based GIS products, such as ArcView, you should be familiar with modal and modeless behavior. For example, the legend editor in ArcView is modal, as will be the legend editor you will build in later chapters.

The *Show* method, as in *Form2.Show,* places a form on the screen. You currently have two forms, *Form1* and *Form2.* To reference form *Form2* in form *Form1*'s code sheet, you have to name the form (*Form2*) followed by a period, then the method. More generally, you access methods and properties found in one form from another form's code sheet using the following syntax.

```
Objectname + a period + method or property name
```

If you simply type *Show* in form *Form1*'s code page, VB would assume you were referring to *Form1.* Because that form would already have to be on the screen (to be able to click on the Command button), this command would have no visible effect.

Let's return to our program. Continue with the following.

22 In form *Form2*'s code page, add the following code.

```
Private Sub Command1_Click()
  Unload Me
End Sub
Private Sub Text1_Change()
  Form1.Text1.Text = UCase(Text1.Text)
End Sub
```

There are some new things in these subs. (Subs are discussed in detail later in the book.) *Unload* is a VB statement that unloads a form or control from memory. You have to tell VB what form or control to unload. The *Me* keyword is used to reference the current form (*Form2*). Thus, when the user clicks on the button in form *Form2,* the form will unload.

The *UCase* function converts a string to uppercase. Look closely at the line that uses this function. On the left side of the assignment operator (the = sign) is *Form1.Text1.Text.* The first part of this statement, *Form1,* tells VB you want to access a control used in form *Form1.* The second part, *Text1,* tells VB that the control you want to access is the text box named *Text1.* The last part of the statement, *Text,* says you want to do something to the *Text* property of the control. The next part of the statement is the assignment operator symbol (=). You might be tempted to call this the equals sign, but it is more accurate to call it the assignment operator. In this example, you are assigning some value to the *Text* property of the *Text1* control found on form *Form1.*

You are assigning the uppercase version of form *Form2*'s *Text1* control's *Text* property. Note that you do not have to have *Form2.Text1.Text,* although that would not hurt. However, because you are in form *Form2*'s code page, the *Form2.* is unnecessary.

23 Run the program.

Your screen should look something like that shown in figure 1-9.

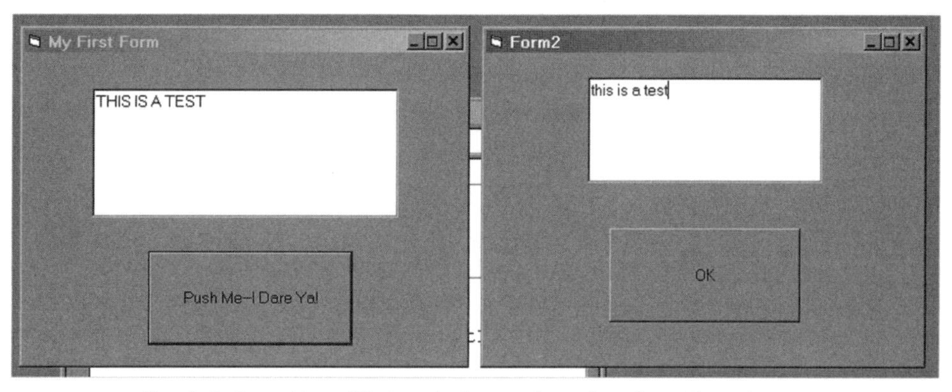

Fig. 1-9. Form Form2*'s text being assigned to form* Form1.

Forms and Controls

Forms and controls are the building blocks of the VB programs you will write. When your program runs, the forms become the windows with which the user interacts. Controls are the tools (for example, command buttons and text boxes) you place on your forms. The forms and controls have properties and methods, and they respond to events.

It is important that you understand the differences among properties, methods, and events. Consider a mapping application. You start with a window showing a map. The map could have many properties, such as the map's extent. The extent is a property. Further, the extent consists of width and height. In addition to extent (and related to it) would be the map scale. These are all map properties. Other properties might be a collection of layers, each with its own properties. The map's methods might include operations that affect the map's display, such as *Refresh* and *FlashShape.*

Now consider events. The following is an example of an event. A user might click on the map, and perhaps drag the mouse. Depending on the map tool used, a different type of event might occur. This might also be set up to zoom in or out on the map. The event in this case is that a rectangle is drawn on the screen.

What happens after an event? An action (method) gets executed. For example, if you know the extent of the rectangle drawn, the map might respond via a zoom method to change its extent. The zoom method will cause a change in properties (the map extent will change), and a redraw of the map will occur. You can think of many mapping events, properties, and methods. Indeed, you will see that MO has many properties and methods. For now, however, let's focus on working with the basic form elements in standard VB.

Forms

This section explores the methods and properties of VB forms.

1. From the VB help index, find the *Form Object* entry and open it. Figure 1-10 shows a form object of VB 6's Help file.

Forms have properties, methods, and events. You will not use all of the possible form properties, methods, and events, but the more important of these are described in the following. The following are form properties.

- *BackColor:* The background color of a form. You have read and write access to this property.
- *BorderStyle:* The border around the form. This can have many styles, such as single line or double line.
- *Caption:* This is the caption that appears in the top stripe of the form. The default name is the form name (e.g., *Form1*), but you can set this to whatever caption you like. In the previous example program, you set the form's caption with the line *Form1.Caption = "Hey–You hit me!"*

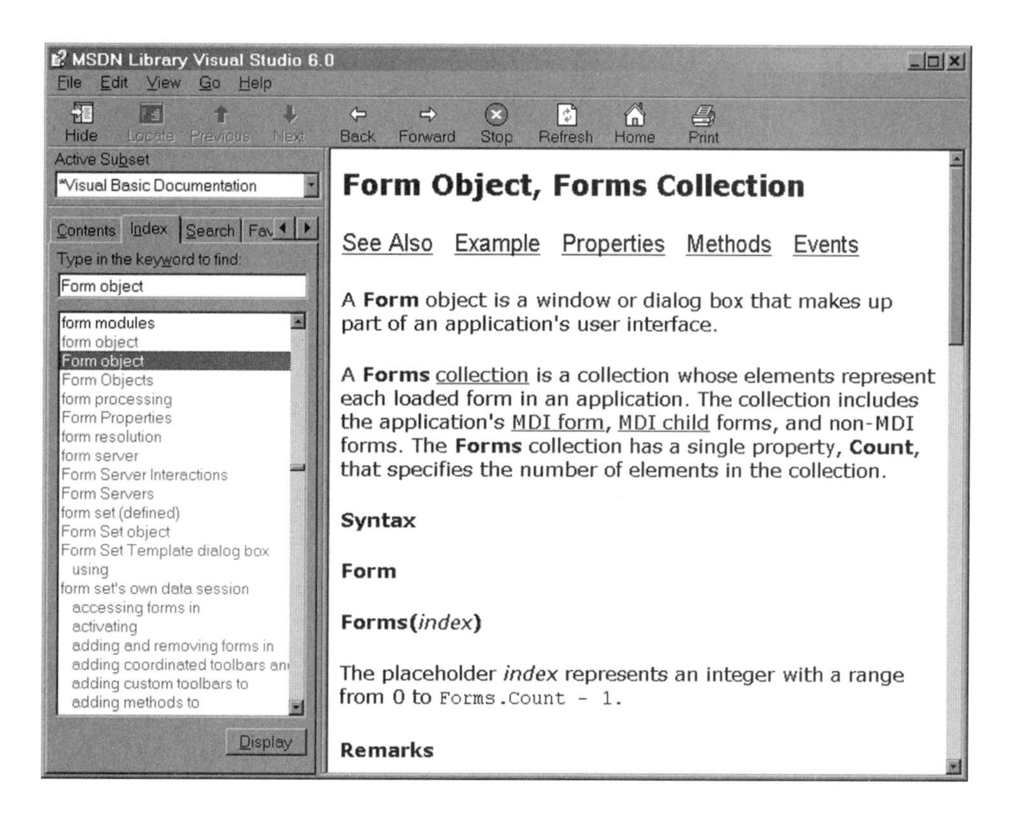

Fig. 1-10. Form object of Visual Basic 6 Help file.

The following are form methods.

- *Show:* Displays the form on the screen.
- *Hide:* Removes the form from the screen.

The following are form events.

- *Load:* Called when the form is loaded into the computer's memory. The load event is typically used to set the starting values of all controls on a form.
- *Unload:* Called when the form is about to close. This event is often used to capture values on the form and assign them to variables. For example, you might use a form to set the title of a map. When the user closes the form, you would save the title they have entered and assign it to the map.

Controls

Let's turn our attention to controls. Controls are the tools you place on a form. There are many controls, such as command buttons, text boxes, list boxes, check boxes, and others. For now, let's consider some properties, events, and methods of command buttons. In material to follow, you will work with some of the other controls.

Command buttons, like the buttons used in the previous examples to trigger form actions, also have properties, methods, and events. The following are some properties you will find useful.

- *Caption:* Sets or gets the caption on the button. You set the caption on buttons in the previous example programs.
- *Name:* The name of the command button. It will be helpful to give controls, like command buttons, logical names (such as *cmdOK* or *cmdCancel*) rather than the default names of *Command1*, *Command2*, and so on.
- *Height and Width:* Use these to set or get the size of the command button on the form.

Perhaps the most important command button events are *Click* and *GotFocus*.

- *Click:* This event occurs whenever the user clicks on the Command button.
- *GotFocus:* This event occurs when the Command button becomes the active control. When a command button has focus, pressing the Enter key is the same as clicking on the control. Related to the *GotFocus* event is the *SetFocus* method.
- *SetFocus:* Use this method to force the Command button to have focus.

The VB Controls Sample Program

VB comes with several sample programs. A particularly helpful program is the Controls program. This program allows you to investigate the behavior of several of the controls you are likely to use. To see how buttons (and other controls) work, let's load the following sample project that comes with VB.

1. Find where the VB sample files are located on your computer or copy them from the Microsoft Developer Network CD (*\Samples\VB98*).

***NOTE**: If you do not have access to the VB samples, just read along.*

2. Go to the Controls subdirectory. Load the project, and open the forms *frmButton* and *frmMain.*

Note how the forms are named. The first three letters are lowercase and indicate the type of object (i.e., *frm*). The next word starts with a capital letter and indicates the control type. Your screen should look like that shown in figure 1-11.

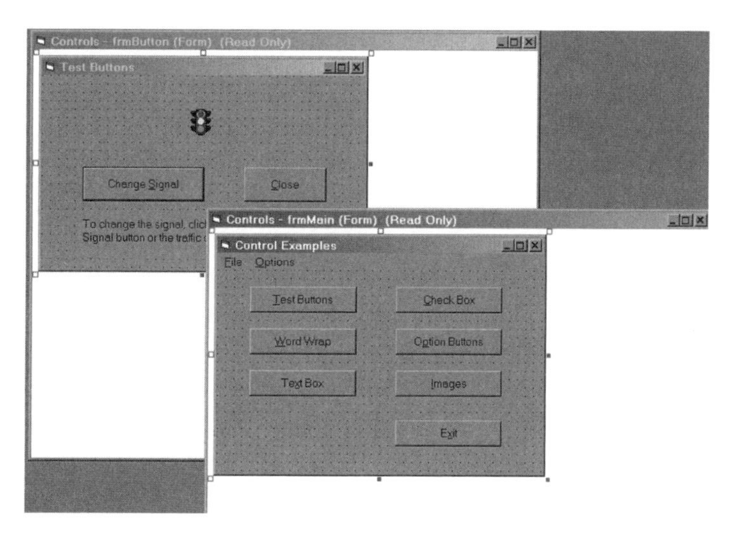

Fig. 1-11. Forms frmButton *and* frmMain.

3. Run the program and use the Test Buttons option.

As you click on the Change Signal button, the traffic light will appear to change from red, to yellow, to green.

4. Stop the program and return to the IDE.

How do you suppose the light colors change? If you click on the properties for the traffic light image, there appears to be no property for determining which bulb is on. How is this done?

5. In the form editor, drag the signal light to the left or right.

You will see that there are three images, one on top of the other. One has the red light highlighted, one the yellow light, and one the green light. The program changes the light by changing the drawing order of the images. Whatever image is on top represents which light is "on." You will use this same strategy when building a legend editor in Chapter 8.

There are two types of text display controls: the Text Box control and the Label control. To see how these controls work, run the Controls sample program, and click on the WordWrap option. The following are properties, methods, and events associated with label boxes. You will find more detailed explanations in the VB Help file.

- *Alignment:* Use this property to set whether text should be left-justified, right-justified, or centered. This also applies to text boxes.
- *AutoSize:* When set to True, this property forces the label box to resize itself (up to the size of the form) to display all the text.
- *Caption:* This property is the text prompt next to the label box.
- *Visible:* This property is used to set or indicate a text display's visibility.
- *WordWrap:* If the AutoSize property is True and this is True, the label box will expand vertically to display the text. If AutoSize is True and WordWrap is False, the label box will expand horizontally.

The following is a label control method.

- *Refresh:* Forces the control to be repainted (rewrites the text in the label control).

The following are text box control properties, methods, and events.

- *Enabled:* When set to True, this property allows the text box to respond to user events. That is, whether the user can type in the text box, highlight text, or click on the text box.

- *Locked:* When set to True, this property precludes the user from editing the text in the text box. When False, the user can edit the text in the text box.

The following is a common *TextBox* method.

- *SetFocus:* Makes the text box the active control.

The following is an important *TextBox* event.

- *Change:* This event occurs whenever the content of the text box changes.

Let's try to determine how the *Change* event occurs.

1. Return to the Controls program and add the following sub to the *frmText* code.

```
Private Sub txtDisplay_Change()
 MsgBox "change is in the wind"
End Sub
```

2. Run the program.
3. Click on the Text Box button. Try to change the content of the leftmost text box.

Every time you press a letter on the keyboard, the message box appears. Suppose you had a text box where the user could enter a map title. Is this the behavior you would want? Probably not. The point is that the *Change* event is fired each time the user types in the text box.

Check Boxes, Radio Buttons, and List and Combo Boxes

We often give users choices. In some cases, choices are mutually exclusive; in others, they are not. For example, you could allow the user to turn map layers on or off. One layer's behavior (on or off) probably would not affect the behavior of another layer. However, if you gave the user a set of map projections from which to choose, you would probably want one and only one map projection used at any time. The Controls sample program contains two options that display how check boxes and radio buttons work. These correspond to the Checkbox and Option Buttons options of the program.

Check Boxes and Radio Buttons

Check boxes can be turned on or off, but they have no effect on any other check boxes. You use the check box value property (i.e., is the check box turned on or off?) to control how your programs work. For example, check boxes are often used to toggle layer visibility in maps. Let's begin by considering check boxes. The most important property of the check box control is its value.

- *Value:* This property is either 1 (the box is checked) or 0 (the box is not checked).

The box value is usually set by the following event.

- *Click:* This event occurs when the user clicks on the check box.

To practice adding a new check box, perform the following steps.

1. Study the check box example in the Controls sample program. Open the form *frmCheck* and its code page (see figure 1-12).

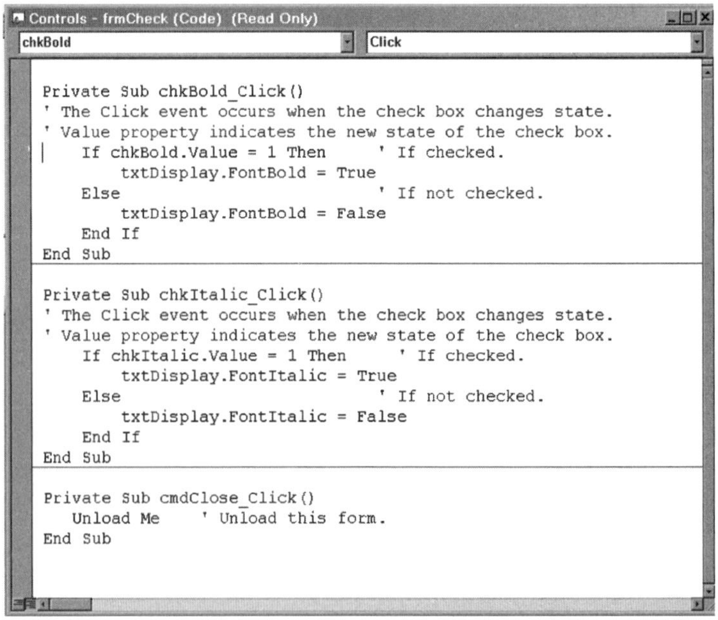

Controls - frmCheck (Code) (Read Only)

chkBold | Click

```
Private Sub chkBold_Click()
' The Click event occurs when the check box changes state.
' Value property indicates the new state of the check box.
    If chkBold.Value = 1 Then     ' If checked.
        txtDisplay.FontBold = True
    Else                          ' If not checked.
        txtDisplay.FontBold = False
    End If
End Sub

Private Sub chkItalic_Click()
' The Click event occurs when the check box changes state.
' Value property indicates the new state of the check box.
    If chkItalic.Value = 1 Then     ' If checked.
        txtDisplay.FontItalic = True
    Else                          ' If not checked.
        txtDisplay.FontItalic = False
    End If
End Sub

Private Sub cmdClose_Click()
   Unload Me    ' Unload this form.
End Sub
```

Fig. 1-12. The frmCheck *form in the Controls sample program.*

2. Add a new check box named *chkUnderline* (for an underlining function) to the *frmCheck* form (see figure 1-13). In the *frmCheck* code page, add the following code.

```
Private Sub chkUnderline_Click()
  If chkUnderline.Value = 1 Then
    txtDisplay.FontUnderline = True
  Else
    txtDisplay.FontUnderline = False
  End If
End Sub
```

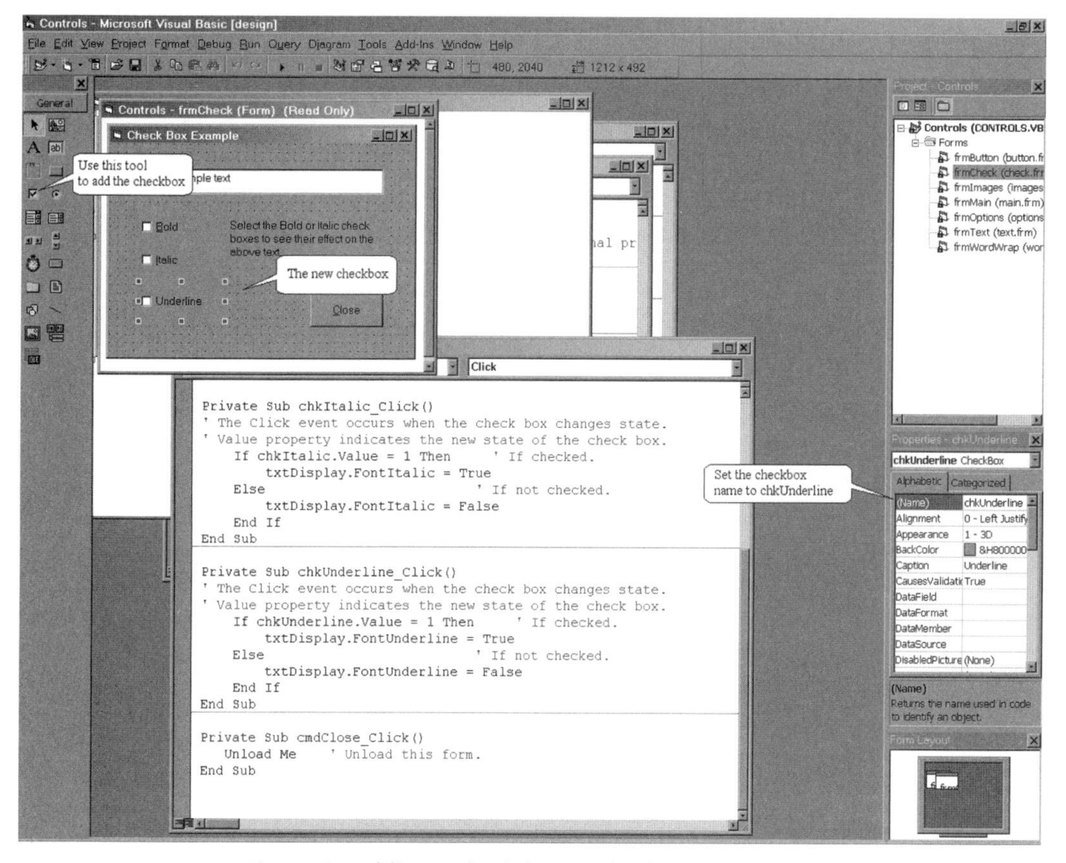

Fig. 1-13. Adding a check box to the form chkForm.

3 Run the program: Click on the Check Box button, and then on the Underline check box.

The text in the window should be underlined.

Radio buttons present a set of two or more options that are mutually exclusive. If you place several buttons on a form, they are considered a group and only one can be chosen. If you want sets of mutually exclusive groups, you put each set in a separate frame.

Consider the combo example form *frmOptions* in the Controls sample program.

1. Resize the operating systems group box (make it very small).

When the frame containing the buttons becomes small, the radio buttons within it seem to disappear. This is because they belong within the frame. Therefore, the frame must be large enough to show them if they are to be seen.

2. Resize the box to its original size.
3. Run the program to see how these options work.

The Operating System radio buttons are mutually exclusive, as are the radio buttons corresponding to the types of processor.

4. Stop the program and return to the *frmOptions* form.
5. Cut the two check boxes out of the operating systems group frame. (It is assumed you know how to cut and paste.)
6. Resize the group frame, making it small again.
7. Paste the two radio buttons outside the operating systems group frame.
8. Run the program.

If you open the options form, all five radio buttons on the form (the two operating system buttons and the three processor buttons) are mutually exclusive. That is, selecting one turns all the others off.

If you resize the frame that contained the operating system radio buttons so that the radio buttons are inside it, the program will still not work. That is, all five buttons (those for operating system and those for processor) will be mutually exclusive. The buttons must be pasted into the frame. The point is, with frames you must paste controls into them to be part of the frame. Simply drawing a frame around a group of controls is not enough.

List and Combo Boxes

List boxes and combo boxes allow users to select items from a list. The main difference is that a combo box allows users to type entries into the list, whereas a list box allows the user to select only from a list of options. The following are important list box properties.

- *ListCount:* The number of items in the list.
- *ListIndex:* The index number of the currently selected item.
- *Multiselect:* Sets or indicates whether the user can make multiple selections from a list. If the value of *multiselect* is 0, the user can select only one item. If the value is 1, multiple items can be selected. If the value is 2, the Shift and Control keys can be used to select multiple items. For example, holding the Shift key down and clicking on an item selects all items from the previously selected item to the current item in the list.
- *Selected:* This property sets or indicates whether an item in the list has been selected.
- *Style:* The meaning of this property is dependent on whether the control in question is a list box or a combo box. List boxes can be standard (a list of items is presented) or check box (each item in the list can be checked on or off).

The following are some useful methods associated with list boxes.

- *AddItem:* Adds a new item to the list to be displayed.
- *Clear:* Removes all items in the list.

The following event applies to the check box style of list box.

- *ItemCheck:* This event is fired when the user clicks on the check box portion of a list box whose items have check boxes associated with them.

Combo boxes combine the list box with the edit box. That is, the user can select an item from a list or type in her own item. Combo boxes have many of the same properties, methods, and events as list boxes. Like edit boxes, combo boxes have a text property for capturing anything the user might enter, as well as a change event to signal when a change occurs.

Summary

This chapter has introduced some aspects of programming, with particular emphasis on objects and object-oriented programming. The first part of the chapter looked at objects in a generic way, using houses, people, and bicycles as examples. In the second part of the chapter, you used the Visual Basic IDE to work with some of VB's objects. The VB projects you developed made use of forms, and you were able to put other objects (such as command buttons, text boxes, and check boxes) on those forms.

Each form is associated with a code page—the place you type in your code. In the second VB project, you saw how forms "communicate." The final sections of the chapter used the VB sample program Controls to illustrate how selected controls work. You saw that check boxes respond to click events, and that the layout of radio buttons determines whether those buttons are in separate groups or all in one group.

Much of this material may be new to you, but it will be reinforced in the coming chapters. When working in later chapters, you may find it helpful to return to the definitions of properties, methods, and events presented in this chapter. Learning a new language requires repetition, and that is what the chapters and projects in this book will give you—repeated experiences in working with the forms and controls introduced here.

Chapter 2

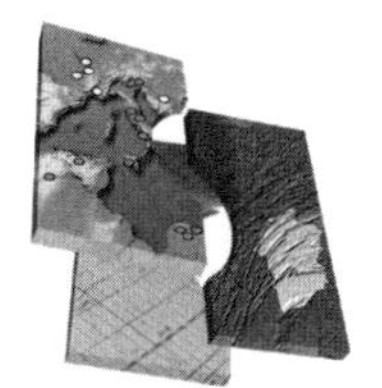

Using ActiveX Controls

Introduction

In this chapter you will learn that it is possible to create powerful programs with very little programming on your part. This is accomplished via one of the latest advances in computer software engineering: components. Components are pieces of existing programming capable of being incorporated into other programs.

Components are often called plug-ins because you can plug them into your program and use them with very little overhead. In this chapter you will work with two components: one an ActiveX object (an OCX), the other an ActiveX DLL (dynamic link library). The differences between OCXs and DLLs are explored later in the book. Here you will see how easy it is to create some interesting applications using components.

A Data Access Application

Let's look at a simple data access application. It was not many years ago that accessing data from different databases was difficult. Now we have ODBC (open database connectivity) and DAO (data access objects). Data access objects are sometimes referred to as ADO, for ActiveX data objects. There are controls you can add to forms that have database access tools built into them. This makes it relatively easy to work with databases.

CD-ROM NOTE: *The VB project in the* Chapter 2_1 *directory on the companion CD-ROM starts here.*

1 Start a new project in VB.

2 To add a new control to your program, click on Project and then on Components in the menu, or simply press the Control and T keys simultaneously (Ctrl-T).

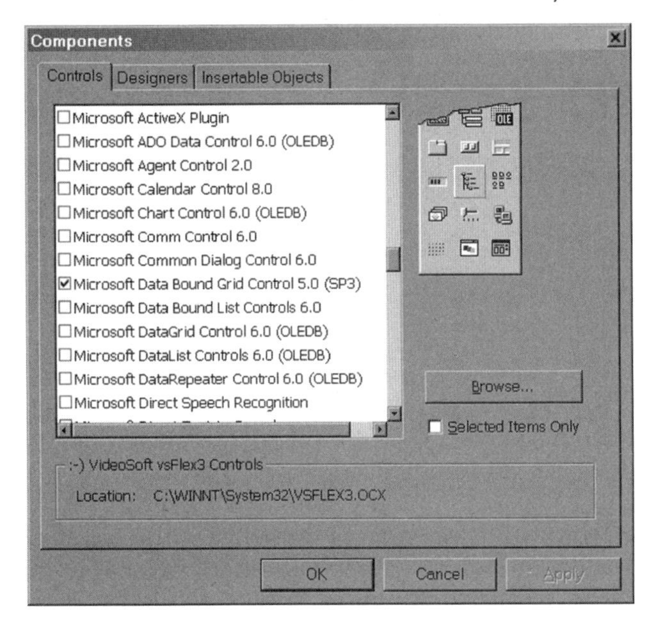

Fig. 2-1. Adding the data bound grid control.

A list of components you can add will appear.

3 Scroll down the list and activate (check) the Microsoft Data Bound Grid Control option (see figure 2-1).

This will add a data control and a *dbgrid* control to the toolbar (see figure 2-2).

4 Create a form with the following elements on it: a data control, a *dbgrid* control, a list box control, and two command buttons.

Your screen should look like that shown in figure 2-3.

5 Set the following properties (use the Properties window for each control):

❒ Set the form caption to Bookstore.

❒ Set the *Data1* database name to *path\Biblio.mdb*, where *path* is the path to the sample data, most likely *C:\Program Files\Microsoft Visual Studio\VB98*. You can use the Open File dialog to find it.

❒ Set the *Data1* record source to All Titles.

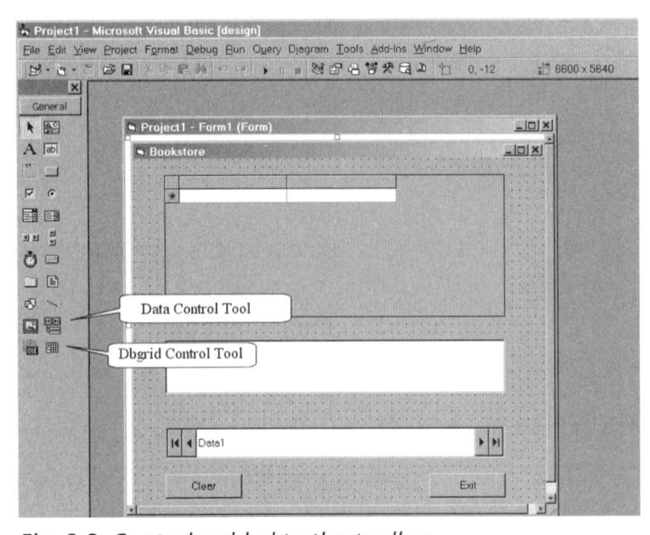

Fig. 2-2. Controls added to the toolbar.

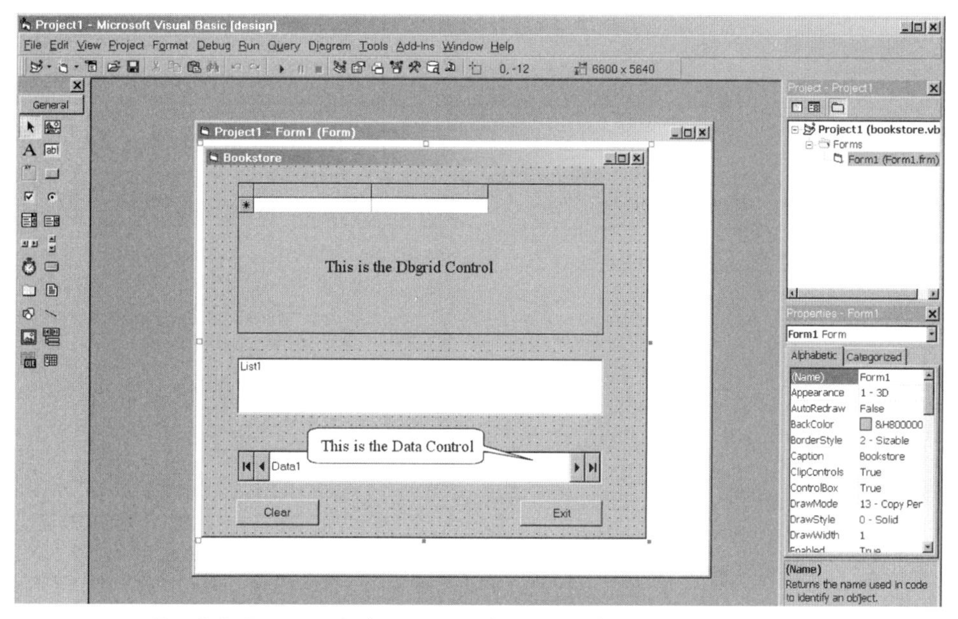

Fig. 2-3. Form and elements to be created.

- ❒ Set the *DBGrid1* data source to Data1.
- ❒ Set the *Command1* name and caption to Clear.
- ❒ Set the *Command2* name and caption to Exit.

6 Open the code window for this form, and add the following subs.

```
Private Sub Clear_Click()
    List1.Clear
End Sub
Private Sub Exit_Click()
    Unload Form1
    End
End Sub
```

If you looked up the items listed in Chapter 1, you should know what these subs do. The first clears all items in the *List1* list box, and the second unloads the form when the user clicks on the Exit button. The *Data1* control is for navigating through the database. That is, when its value changes (i.e., the slider moves left or right), you will reposition the pointer to the database to the current title.

```
Private Sub Data1_Reposition()
   Data1.Caption = Data1.Recordset("Title")
End Sub
```

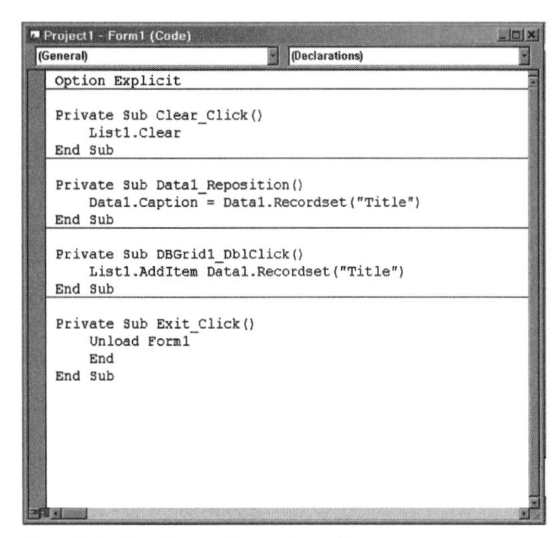

Fig. 2-4. The form Form1 code page.

Because the *dbGrid1*'s data source is *Data1*, whenever you reposition the pointer in the database, the grid value is automatically updated. You can read more about Data Bound grid controls by going to "Data-Bound Controls, Using in Visual Basic" in the VB Help file. You will see that VB has an object type called a *record set*. You will see later that MO also has a *recordset* object. Finally, if the user double clicks on an entry in the grid, you will add the entry to the list (as if you were building a list of books to buy).

```
Private Sub DBGrid1_DblClick()
   List1.AddItem Data1.Recordset("Title")
End Sub
```

Your code page should look like that shown in figure 2-4.

7 Run the program (see figure 2-5).

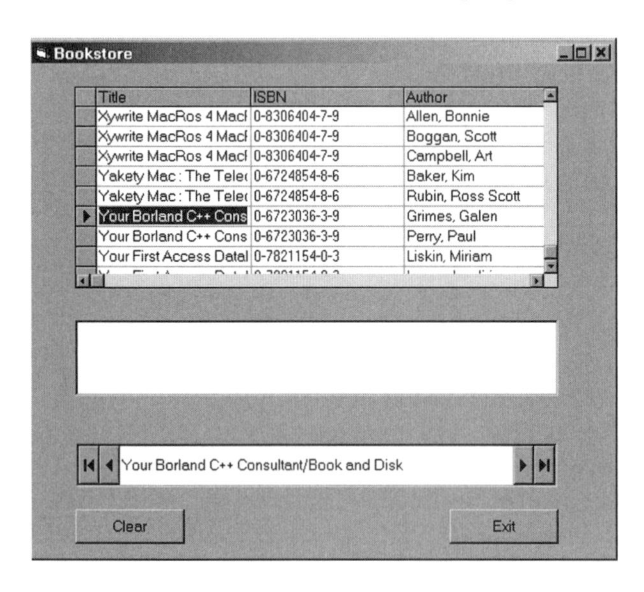

Fig. 2-5. Result of running previous sub.

Notice that you can resize the columns in the *dbgrid*, scroll through the elements in the database using the data control, and select an entire row or column from the *dbgrid*. How much code did you have to write to enable these methods? None! This is the beauty of ActiveX controls.

ActiveX control methods are accessible with very little programming overhead. This removes much of the drudgery when writing code for an application. You do not have to tell the control what a scroll bar or a column resize arrow does; those behaviors are already programmed into the control. It also knows that if more titles are entered into the list box than space allows, a scroll bar should be added. This type of built-in functionality saves you a lot of work. Do not be fooled, however; life in the ActiveX world still requires some effort.

Think a minute. You have just created a custom interface to an Access database and you only had to write 13 lines of code!

Adding DLL Controls

As you work with MO, you will learn to work with an OCX, or ActiveX, control. There are, however, other types of controls you can work with. For example, you can add certain DLLs (not all DLLs can be added via a toolbar, but some can). Let's try one now.

NOTE: *Before starting this project, make sure Windows Explorer (or Windows NT Explorer) is configured to show all files. To do this, start Windows Explorer, click on View, and then click on Options. Make sure Show All Files is active (checked), and that Hide System Files is not active (checked).*

CD-ROM NOTE: *The VB project in the* Chapter 2_2 *directory on the companion CD-ROM starts here.*

1. Start a new project.
2. Add a control (Ctrl-T, or click on Project > Components in the menu bar).
3. On the Controls dialog, click on Browse.
4. Go to *C:\Windows\System* (if you are using Windows NT, this is probably located in the *Winnt\system32* directory). Set the file type to All Files.
5. Select the DLL file *shdocvw.dll* (see figure 2-6).

This adds the Microsoft Internet controls to the list of available controls.

6. Add the control to the toolbox (click its check box on).
7. On the form, add the Internet control (use the globe icon in the toolbar), a text box named *txtLocation,* a command button named *cmdSurf* (with the caption *Surf*), and two label boxes, *lblStatus* and *lblProgress.*

Your form should look like that shown in figure 2-7.

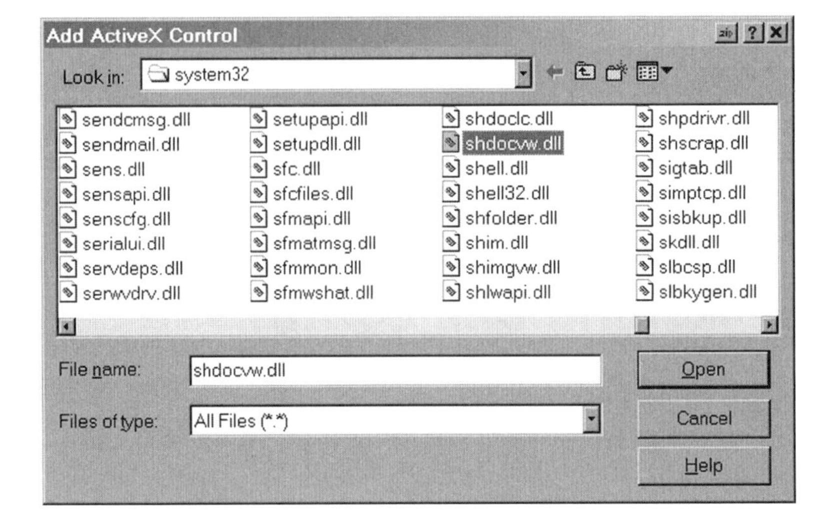

Fig. 2-6. The DLL file shdocvw.dll.

8 For *txtLocation*, set the text to your favorite web site, such as *http://web.utk.edu/~utkgeog*.

9 Add the following subroutines to the form's code page. Do not worry if you do not understand all code details. Eventually this will make sense.

TIP: *Remember, you do not have to type the name of the sub. You can choose it by selecting the object and method from the drop-down lists at the top of the form's code window.*

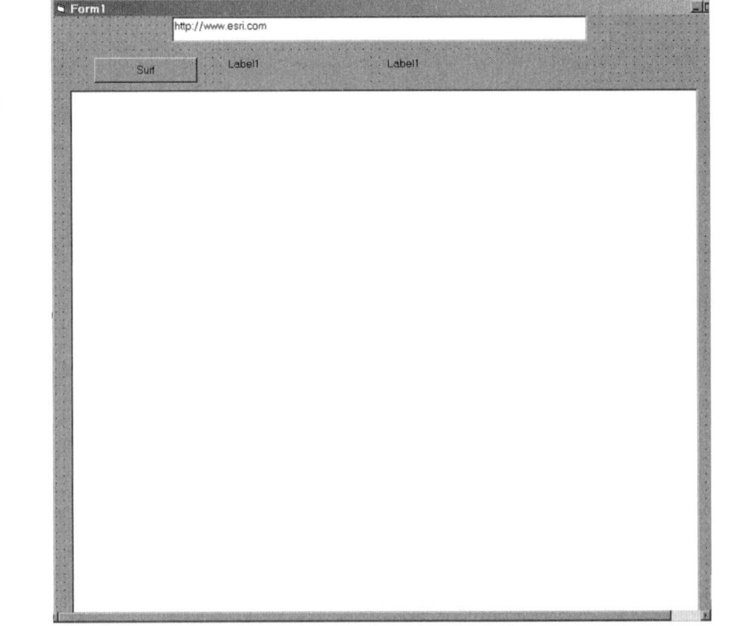

Fig. 2-7. Components added to the form.

```
Private Sub cmdSurf_Click()
  If txtLocation <> "" Then
    WebBrowser1.Navigate txtLocation
  Else
    txtLocation.SetFocus
    Beep
  End If
End Sub
Private Sub Form_Initialize()
   With WebBrowser1
    If txtLocation <> "" Then
       .Navigate txtLocation
    End If
  End With
 End Sub
Private Sub txtLocation_GotFocus()
  cmdSurf.Default = True
End Sub
Private Sub WebBrowser1_BeforeNavigate2(ByVal pDisp As Object, URL _
    As Variant, Flags As Variant, TargetFrameName As Variant, _
    PostData As Variant, Headers As Variant, Cancel As Boolean)
  With WebBrowser1
    If .Busy Then .Stop
  End With
End Sub
Private Sub WebBrowser1_DownloadBegin()
  lblProgress = " Downloading..."
End Sub
Private Sub WebBrowser1_DownloadComplete()
  lblProgress = "Done"
End Sub
Private Sub WebBrowser1_NavigateComplete2(ByVal pDisp As Object, _
    URL As Variant)
  txtLocation = URL
End Sub
Private Sub WebBrowser1_StatusTextChange(ByVal Text As String)
  lblStatus = Text
End Sub
Private Sub WebBrowser1_TitleChange(ByVal Text As String)
  Me.Caption = "The World Wide Web - " & WebBrowser1.LocationName
End Sub
```

10 Run the program.

11 If you are connected to the Web, enter a new URL in the text box (for example, *www.gistools.com)* and click on Surf.

Obviously there are more controls you could put in this example. For example, forward and back buttons would be helpful. Nonetheless, you have now built your own web browser!

A Closer Look at VB Projects

When you build VB programs, you create and combine many types of program elements, such as forms, classes, modules, resources, and ActiveX controls. All elements are tracked in a project file. (The file extension for a project file is *vbp*.) Forms are placed in a form file (*frm* extension), one for each form. The properties of the controls on the from are kept in a file with an *frx* extension (a binary file).

You might also employ class modules (*cls*). You will see that class modules store information about the objects you create. Eventually, when you create an executable file (an *exe* file), the various parts of your project (the forms, the class modules, and possibly other files) get written into the *exe* file, and that is what you distribute. Distributing programs is not as easy as in the days of DOS. Because windowing environments use components, you must make sure your program has all components it needs.

Summary

In this chapter you used an ActiveX control and a dynamic link library to build two powerful programs. The data bound grid control allowed you to develop a database-browsing program with only 13 lines of code. The second program used a DLL that contained functions for surfing the Web. You used standard VB controls (such as the text box, in which a user would enter the URL) and the Surf command button to provide an interface for the user to access the DLL's Internet-enabled functions.

The first two chapters have covered a lot of material. You might think (and justifiably so) "I just type what I am told to type, but there is a lot going on that I don't understand." Now that you can see some of the power you can build into your programs, it is time to go back and fill in some details. In the next chapter, you will look at some programming basics, including variable types, just what is meant by the keywords *Sub* and *Function*, and so on.

Chapter 3

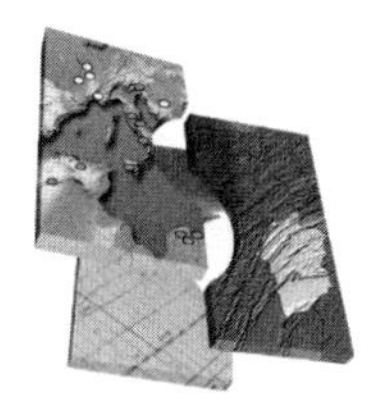

Programming Basics

Introduction

It may seem that we have the cart before the horse, discussing programming basics after we have written some programs. However, having some exposure to VB and a sense of how programs work will make this introduction to programming that much easier. Therefore, now that you have jumped into the deep end of the pool, you can begin to explore programming basics. If some of this material seems a bit abstract at this point, do not worry. This material is expanded upon throughout the remaining chapters. The ideas presented here should become second nature.

Types of Code Modules in VB

VB programs usually consist of a set of forms. Each form has controls. For each form there is a code file (called an *.frm* file) that corresponds to items on the form.

Each form's code file contains VB subs for handling events. Events can be mouse clicks, drags and drops, and so on. You can see what events can affect a form or control by clicking on the two drop-down lists at the top of the code-editing window (figure 3-1).

At first, all your programs may need is the code associated with a form. However, as programs become more robust, you might have some code that needs to be accessed from several forms or controls (or more accurately, from objects). For example, in a mapping program you might need some code to handle certain functions, such as translating colors from a palette list to colors (usually hex numbers) used by the computer.

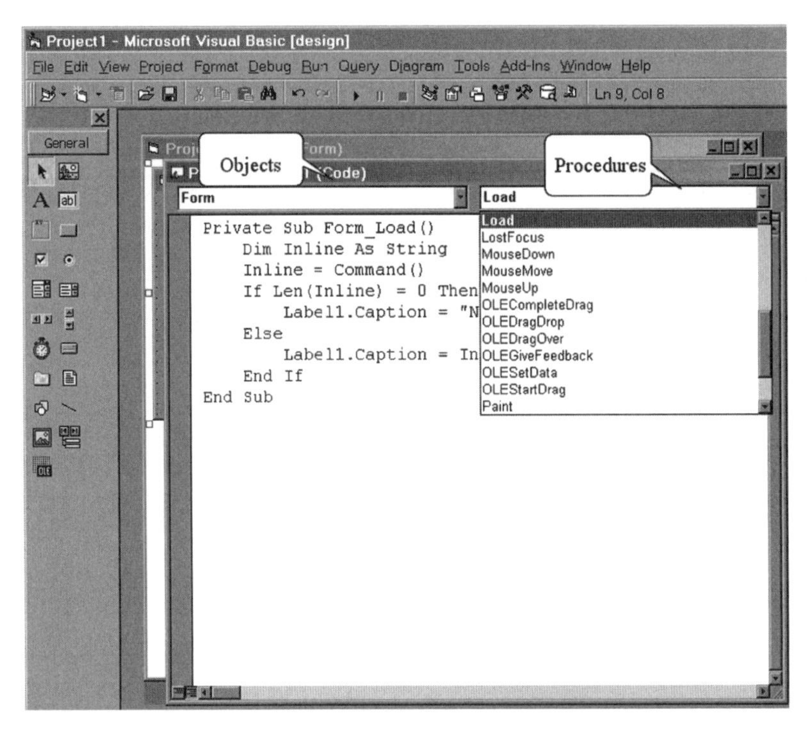

Fig. 3-1. The Objects and Procedures drop-down lists.

You might have some general functions for dealing with selected features. The selected features might be selected from a map, from a table, or via a query. In any case, both the map and its corresponding table will have to be updated. It makes sense to put these functions in a module all other modules can read. Such functions are placed in standard modules (these files have an extension of *.bas*). Code in a standard module can easily be called by all other modules. The reason for the *easily* is that you can call one form's procedures from another form, but there is a bit of overhead involved. In Chapter 1, you assigned text to a text box in form *Form1* from a function in form *Form2*. To do that you had to explicitly reference form *Form1* and its controls when accessing them in form *Form2*. This was accomplished by the following line.

```
Form1.Text1.Text = UCase(Text1.Text)
```

This line appeared in form *Form2*'s code page, and therefore the *Text1* control in the statement *UCase(Text1.Text)* did not require an explicit reference to form *Form2*. However, the *Text1* control on the left side of the assignment operator is from form *Form1*.

Therefore, referencing it required the additional overhead of adding the form *Form1* prefix, as in the following.

```
Form1.Text1.Text
```

Forms and controls are objects built into VB. However, there may be instances where you want to create your own custom objects. To do this, you create class modules (**.cls*). In the class modules, you define the properties and methods of the classes you construct. You will use such constructions in Chapter 12, in conjunction with managing selected map features.

Examine figure 3-1. Where do the names of the functions in the right-hand list box come from? These functions correspond to events VB recognizes, such as a form load or a mouse click or drag. Although Windows-based programming languages provide function headers for events that can take place in windows, they do not tell you what should happen when those events take place. It is up to you to write code (if it is needed) to indicate how your program should respond to those events. It is likely there will be many events your program can ignore. However, there will be a few you must handle.

Looking at VB, then, it seems that much of the structure of a program is predetermined. To an extent this is true. However, it is not the end to program design. The most important thing you can do before starting to code is to have a clear idea of how your program should look and behave. That is, you need to *make some design decisions before writing code.* It is difficult to make good time on a trip when you do not know where you are going! In classes and textbooks this is often overlooked because you are given particular tasks to complete. However, when you get to a major project, you will need to design your own "assignments."

■■ The Mechanics of Editing Code

Examine figure 3-1 again. The code-editing window lists objects and procedures. However, these are not the only procedures that can be placed in a module. You can create your own procedures. This is often necessary when an object's procedure must call another function or procedure. (You will see more of this later.)

As you type in code, you can enter an object. When you type the period after the object name, all of the object's properties and methods appear (see figure 3-2). This feature of VB is called AutoList Members. As you type the letters of the method or property, the list will highlight the possible candidates. To have the current highlighted candidate typed for you, press the Tab key.

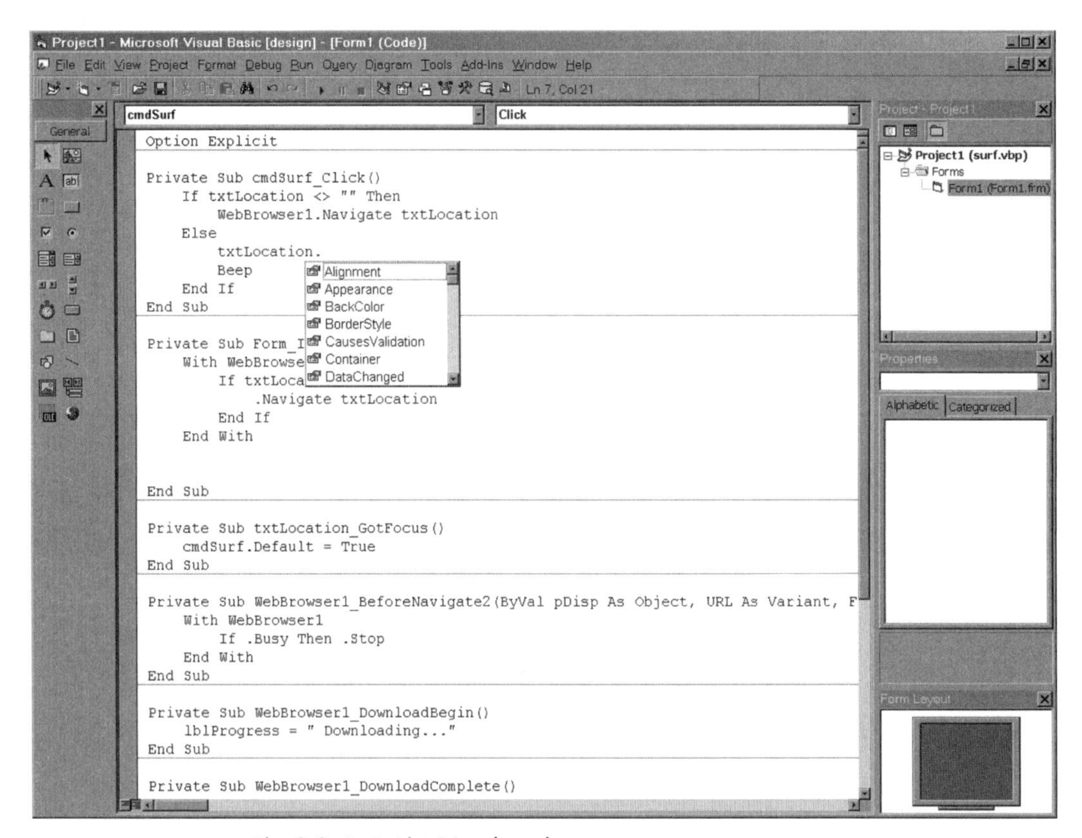

Fig. 3-2. AutoList Members box.

A similar feature is Auto Quick Info, which lists the arguments of a function. For example, try typing *MsgBox*. An information box will appear, which indicates the arguments of the function (see figure 3-3).

VB expects commands to be on a single line. If a line is getting too long, you can continue it by ending the current line with a space followed by an underscore. Nothing else can follow the underscore, not even a comment.

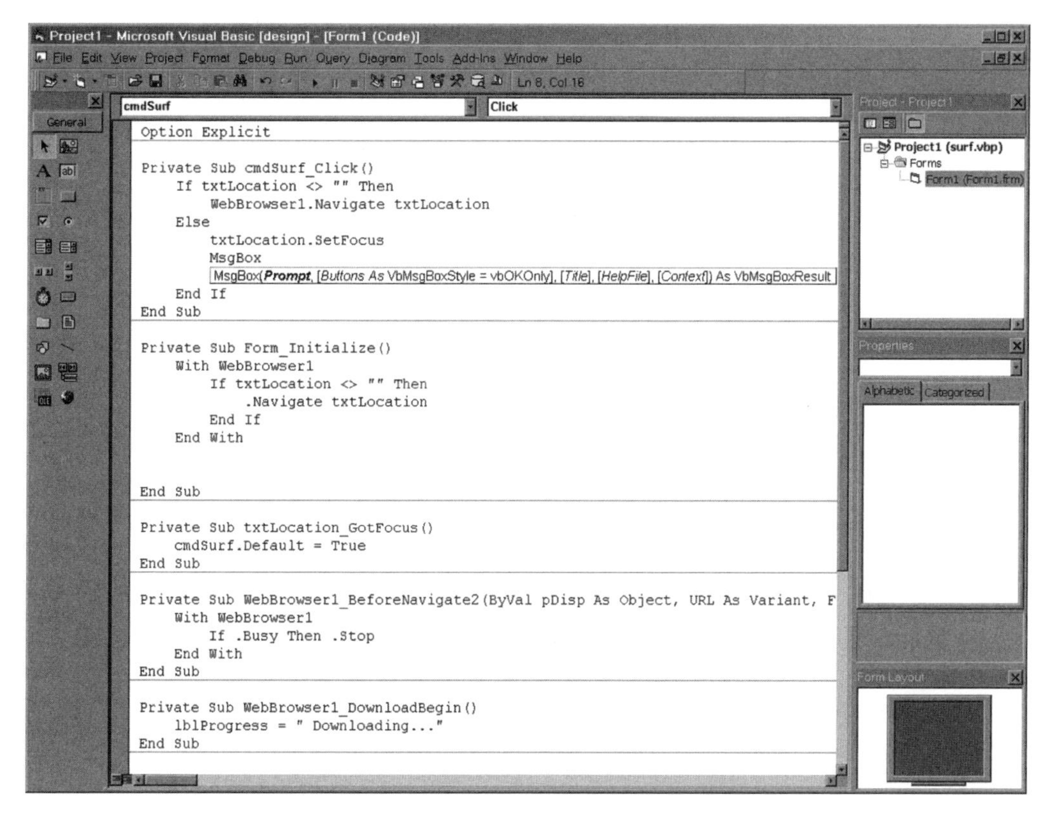

Fig. 3-3. AutoQuick Info and comments.

This is a good time to formally define comments. Comments are lines within code that are not code. That is, VB ignores them. They are neither executed nor interpreted by VB. Comments are typically placed in programs to explain what the code is doing. A comment is indicated by a single apostrophe at its start. Everything after the apostrophe is ignored by VB. In the IDE, comments are displayed in green. Comments are also used to test code. That is, sometimes you might want to try some code, and then "comment it out" to see what happens. That is, place apostrophes before the lines of code so that these lines do not get executed. This is helpful for testing alternative ways of doing things.

Variables

When writing code, you will need to use variables. A variable must have a name, and the name must conform to certain rules: it must begin with a letter, not contain any periods, comply with length restrictions, and not be a keyword. For example, you cannot name a variable *For* or *If* because these are reserved keywords.

A variable is really an address. It is an alias for an address in memory. It makes more sense to say "cost = 25" than to say "memory address 0xFFF3 = 25." When you create a variable, you are essentially grabbing a piece of memory "real estate" and giving it an easily remembered name.

There are two methods for creating variables in VB: implicit declaration and explicit declaration. In implicit declaration, you simply use a variable and VB determines what it is. That is, you do not have to declare it (or dimension it) ahead of time. (If you have worked with Avenue in ArcView, you know that this is how that language works.) In explicit declaration, you tell VB the name and type of every variable used. It may seem that implicit declaration is easier, wherein you simply make things up as you go. This is true, *but implicit declaration is much more dangerous than explicit declaration.* Consider the following.

A list box control has a property called *ListIndex.* You would typically reference this property using a statement such as

```
List1.ListIndex = 1
```

This would set the currently selected item in the *List1* list box to its second item. Suppose, however, you followed implicit declaration and mistakenly typed

```
ListIndex = 1
```

VB would look at this and think, "Aha, the programmer wants to use a new variable called *ListIndex.*" In this case, VB would not think there is any error, yet the program will not do what you intend. That is, it will not set the selected item in the *List1* list box.

To avoid this, you tell VB to enforce explicit declaration. That is, when you start a new project, you use the command sequence Tools > Options > Editor > Require Variable Declaration. This will place the line *Option Explicit* at the top of every code module. If you had done this in the previous example, the line *Listindex = 1* would have caused an error.

Once you have established this editing option, you must declare all variables (explicitly). The general form for declaring variables is:

```
Dim variable name As variable type
```

Here, *variable name* is the name assigned to the variable, and *variable type* is the type of variable. The *As* variable type is not required, but is recommended. (The reason for this is discussed in material on variable types, which follows.) In addition to the keyword *Dim*, you can declare variables with the keywords *Public* and *Private*, discussed in the next section.

Variable Scope

Where and how you declare a variable defines its scope. That is, depending on where you place a *Dim*, *Public*, or *Private* statement, the scope of that particular variable is different for each position. Consider the following code, taken from the VCR sample program that comes with VB. Do not expect to understand what every line does. The focus here is on where variables are declared (the portions indicated in bold).

```
'**********************************************
' Purpose: General module for the VCR sample
' application. Contains shared procedures
'**********************************************
Option Explicit
' Instantiate the recorder class
Public Recorder As New clsRecorder
'**********************************************
' Purpose: Enables or disables buttons on the
'   VCR form based on the current mode.
' Inputs: Button: the Command button calling
'   the procedure.
'**********************************************
Sub ButtonManager(Button As Control)
 Dim vntControl As Variant ' Control value
  ' determine which function button was pushed
  ' and update all buttons and Recorder class
```

The line *Public Recorder As New clsRecorder* is at the top of the file, not within a sub or function, so that the variable *Recorder* is defined throughout the module. Further, because *Recorder* is declared with the keyword *Public*, all modules in the project have access to this variable.

Now consider the line *Dim vntControl As Variant.* This is defined within a sub, and is said to be local or to have procedure-level scope. This means that no other part of this module or this program can read the value created and stored in this sub (*ButtonManager*). Further, because this is a local variable, once the sub finishes executing, the variable *vntControl* ceases to exist. (More accurately, its memory location can be used by some other part of the program.)

Suppose now that the first case reads *Private Recorder As New clsRecorder.* In this case, the keyword *Private* would indicate that only those subs and functions in this module have access to the variable *Recorder.* Other modules in this project could not access the variable *Recorder.* Therefore, there are three levels of variable scope:

- *Local (or procedure) level.* These variables are always private, and are usually declared with a *Dim* statement. As soon as their procedure ends, they disappear.
- *Module level (not defined within a procedure or function) with Private.* These are visible only to subs and functions in the current module.
- *Module level with Public.* These variables are global. All subs and functions in all modules have access to these variables.

Suppose you have a local variable in a sub whose content you do not want destroyed after the sub is executed. In this case, you can define the local variable with the keyword *Static,* as in the following.

```
Static curBankBalance as Currency
```

A static local variable has a lifetime equal to the time your application runs. That is, static local variables are born when the program starts and die when the program ends. A regular local variable is born when its procedure starts and dies when it ends.

Data Types

Previously you learned that a variable name is really an alias for a memory address. You can think of the computer's memory as a large apartment building. Within this building, each variable occupies an address. However, just as different people have different housing needs, different variable types have different storage needs. For example, it is intuitive that storing a number such as 32.3409438230 takes a different amount of space than storing a number such as 2 or 3. The sections that follow explore various data types and their storage parameters.

Numeric Variables

Numeric variables are used to store numbers. Different types of numbers require different amounts of memory, and have different valid value ranges. The amount of space needed by a variable type and the range of values it can take are given in the VB Help file's section "Data Type Summary" (see figure 3-4).

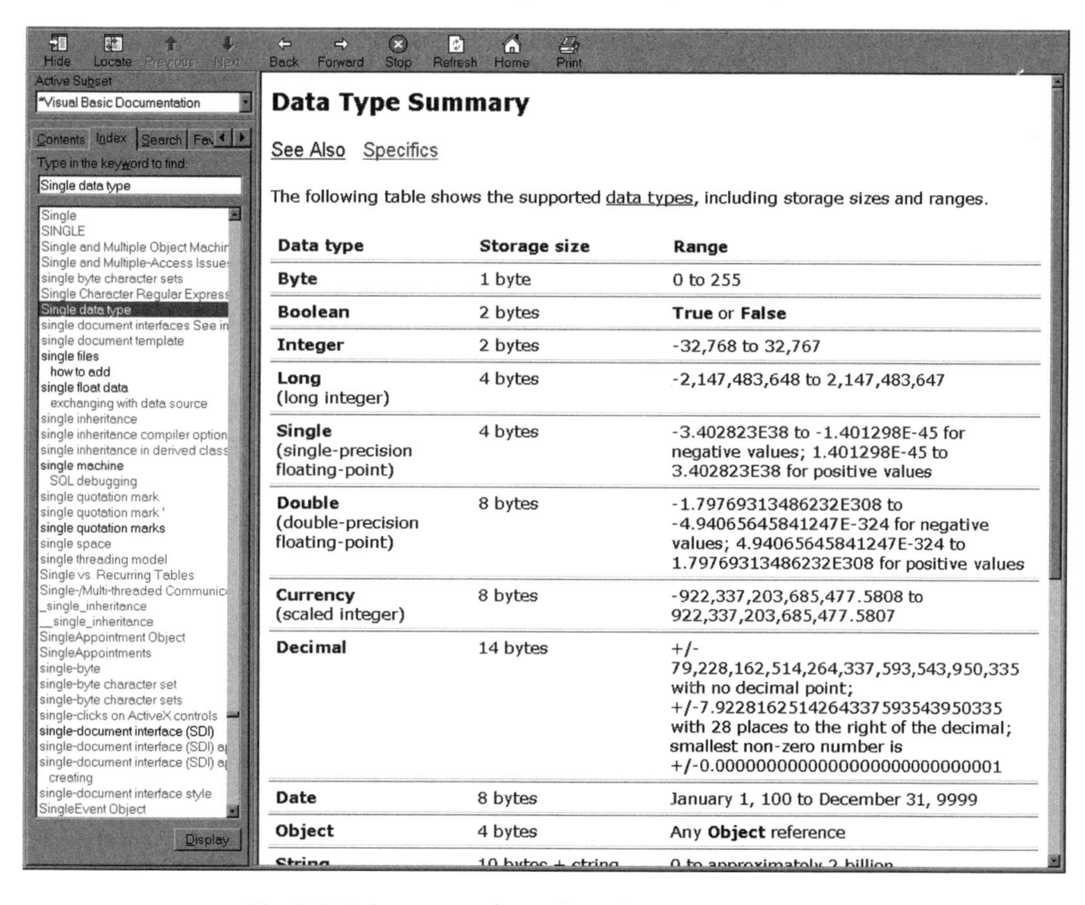

Data Type Summary

See Also Specifics

The following table shows the supported data types, including storage sizes and ranges.

Data type	Storage size	Range
Byte	1 byte	0 to 255
Boolean	2 bytes	**True** or **False**
Integer	2 bytes	-32,768 to 32,767
Long (long integer)	4 bytes	-2,147,483,648 to 2,147,483,647
Single (single-precision floating-point)	4 bytes	-3.402823E38 to -1.401298E-45 for negative values; 1.401298E-45 to 3.402823E38 for positive values
Double (double-precision floating-point)	8 bytes	-1.79769313486232E308 to -4.94065645841247E-324 for negative values; 4.94065645841247E-324 to 1.79769313486232E308 for positive values
Currency (scaled integer)	8 bytes	-922,337,203,685,477.5808 to 922,337,203,685,477.5807
Decimal	14 bytes	+/- 79,228,162,514,264,337,593,543,950,335 with no decimal point; +/-7.9228162514264337593543950335 with 28 places to the right of the decimal; smallest non-zero number is +/-0.0000000000000000000000000001
Date	8 bytes	January 1, 100 to December 31, 9999
Object	4 bytes	Any **Object** reference
String	10 bytes + string	0 to approximately 2 billion

Fig. 3-4. Value ranges for various data types.

Strings

A *string* is a special type of variable. It is actually an array of characters. There are many extremely useful string functions. The following are those you will use in subsequent chapters.

- *Left(thestring, n):* Returns a new string consisting of the *n* leftmost characters of the variable *thestring*.
- *Right(thestring, n):* Returns a new string consisting of the *n* rightmost characters of the variable *thestring*.
- *LCase(thestring):* Converts *thestring* to all lowercase characters.
- *UCase(thestring):* Converts *thestring* to all uppercase characters.
- *Mid(thestring, m,n):* Returns a new string consisting of *n* characters in *thestring*, starting at position *m*.
- *Ltrim:* Removes any leading blank spaces from a string.
- *Rtrim:* Removes any trailing blank spaces form a string.
- *InStr(string1, string2):* Returns the first position where *string2* occurs in *string1*.
- *Len:* Returns the length of the string.
- *Strcomp(string1, string2):* Returns 0 if *string1* = *string2*. It returns –1 if *string1* occurs before *string2* when the strings are sorted lexicographically. Otherwise, it returns 1.

NOTE: *Lexicographic sorts are similar to alphabetical sorts, except that lexicographic sorts handle nonalphabetic values, such as numbers. The following list is in lexicographic order.*

Text

Text1

Text11

Text2

Booleans

A *Boolean* data type can take only two values: True and False.

Variants

Another data type is the *variant* data type. Variants are the "Tupperware" of variables. They can contain anything. That is, a variant can store any type of variable. When you need to use a variant, VB will convert it to whatever data type is needed. It would seem that the easiest thing to do would be to store everything as a variant.

However, you would not want to do this because every time you use a variant variable, VB has to check its type, convert it, and then use it. This greatly slows down execution. This process also consumes a lot of memory. The most appropriate use of a variant is when you are unsure what type of object is being returned from an operation or if a particular procedure requires that the variables you send are variants.

Null Values

Another import aspect of data storage is the null value. You assign a *Null* to a variable with a statement such as:

```
current = NULL
```

You can test for a variable being null with a statement such as:

```
If IsNull(current) then...
```

There is an aspect of null values that can be difficult. If an expression has a null value in it, the result of the expression is null. For example, suppose you had a statement such as:

```
total_packages = inboundAM + inboundPM + outboundAM + outboundPM
```

Suppose also that the values of the variables on the right were:

```
inboundAM = 100
inboundPM = null
outboundAM = 300
outboundPM = 200
```

It would seem that total packages should be 600, but VB will return a value of null!

Arrays

Arrays are another type of variable declaration. Arrays are contiguous chunks of memory used to store variables that have a common name and can be accessed via an index. You are probably familiar with them. Arrays come in two flavors: fixed size and dynamic. Fixed-size arrays have a known maximum number of elements, as in the following statement.

```
Dim Totals(19) as Single
```

This statement declares an array of single-precision floating point values. As you saw from figure 3-4, single-precision floating point numbers (usually referred to as singles) take 4 bytes of memory and can store numbers with up to 38 positions before the decimal point and 45 positions after it.

The value 19 indicates that there are 20 elements in the array. That is, you start counting arrays from 0. You can change the lower bound, but this is not recommended, as most modern programming languages start counting array elements from zero. There are also multidimensional arrays, as in:

```
Dim Matrix(9,14)
```

This statement creates a 2D array of size 10 x 15. Dynamic arrays can be resized at any time. To initialize a dynamic array, use the following statement.

```
Dim Matrix()
```

Once you know the size of the array you need, you can re-dimension it, as in:

```
ReDim Matrix(9,14)
```

However, a *ReDim* statement by itself will wipe out the current content of the array! You can preserve the current content by using the keyword *Preserve* in the *ReDim* statement, as in:

```
ReDim Preserve Vector(19)
```

If you have a multidimensional array, only the upper bound of the last dimension can be changed when using the *Preserve* keyword, as in:

```
Dim Matirx(9, 14)
ReDim Preserve Matrix(9, 20)
```

Procedures

You will typically work with two types of procedures: sub and function. Sub procedures (which you have seen) do not return a value. A function procedure does return a value.

Subs

The syntax for a sub is as follows.

```
{Private|Public|Static} Sub procedurename (arguments)
      things to do
End Sub
```

Every control you place on a form can have subs associated with it. These subs are normally invoked by events, such as a mouse click, but they also can be invoked by calling them from code, as you will see in later chapters. Consider the following sub, which you developed in Chapter 2.

```
Private Sub cmdSurf_Click()
 If txtLocation <> "" Then
  WebBrowser1.Navigate txtLocation
 Else
  txtLocation.SetFocus
  Beep
 End If
End Sub
```

This sub (procedure name *cmdSurf_Click*) is private and takes no arguments. It is invoked whenever the user clicks on the *cmdSurf* button. If you want to terminate a sub, use the *Exit Sub* command.

Functions

Functions are procedures that return a value. For example, you might have a function called *Cube*, as in the following.

```
Function Cube(x as single) As Single
      Cube = x * x * x
End Function
```

If you want to terminate a function, use the *Exit Function* command. Note that the value that is returned is set equal to the name of the function. The last line executed in a function must have the function's name set equal to some value. This is how the value is returned to the calling procedure. You could then call the function from some other sub, as in the following example.

```
Sub cmdCubeButton_Click()
   Dim A As Single
   Dim B As Single
   A = 37.5
   B = Cube(A)
End Sub
```

Note the statement *B = Cube(A)*. Because *Cube* is a function, it returns a value, and therefore it typically appears as the right-hand side of a statement. If you called the function as follows, the function would work fine.

```
Cube(A)
```

However, the result of the function would be thrown away because you did not assign it to a variable. Subs do not return values, and are therefore invoked by standalone calls, such as the following.

```
Sub blah_blah()
   'some statements
   RefreshButtons 'This is a call to another Sub
End Sub
```

To call a procedure in another form's code, just add the name of the form to the front of the procedure name. (The procedure must not be *Private*.) For example, if form *Form1* had a function named *MapRescale()*, you could call it from another form's code (or any other code in the project) as follows.

```
Form1.MapRescale()
```

A similar method can be used for procedures in standard modules.

```
Module1.MapRescale()
```

If *MapRescale* were declared as *Public*, you could simply have

```
MapRescale()
```

Procedures in class modules are different. A class contains the properties and methods of an object. Therefore, to call a method of an object, you must first have an instance of that object. The following is an example.

```
Dim SelSet as New Recordset
Set SelSet = gSelection
SelSet.ClearSelection
```

In this example, you first create an instance of the object (*SelSet as New Recordset*). Next, you use a *Set* command to set the object equal to some value (*gSelection* is a set of selected records in this example). You can then call one of its methods (*SelSet.ClearSelection*).

Passing Arguments to Procedures

Subs, functions, and object methods can take arguments. For example, the *Cube* function took a single value (x) as an argument. There are two ways to pass arguments. The default method is called *pass by reference.* When you pass by reference, you send the memory address of a variable. The function or sub you call can change the value in that address. Consider the following code.

```
Sub Increment(x as integer)
  x = x + 1
End Sub
Sub Test
  Dim A As Integer
  A = 1
  MsgBox A
  Increment(A)
  MsgBox A
End Sub
```

Because *Increment* is being passed a value by reference, the first message box would report a 1, and the second would report a 2. Now suppose you defined *Increment* as follows.

```
Sub Increment(ByVal x as Integer)
```

When you *pass by value,* the receiving program (*Increment*) gets a *copy* of the variable, and therefore the original value does not change. In this case, both message boxes will return a value of 1. There may be instances in which you want this type of behavior. The choice of passing by reference or passing by value can make a big difference in how programs behave.

VB allows you to incorporate optional arguments in code. If optional arguments are not passed, they are not used. For example, the *MsgBox* function must have a prompt sent to it. However, other arguments (such as those used for buttons or a title) are optional. Procedures can also have default values. In the *MsgBox* function, the default value of buttons is 0 (that is, a message box with no buttons on it).

Finally, you can use named arguments. Named arguments allow you to list the parameters passed to a function in any order. Suppose a mapping procedure named *AddLayer* required you to send it a layer name, a layer type, a color, and a symbol. Even if the procedure definition specified the arguments in this order, you could still call the function using the following.

```
AddLayer type:=poly, name:="Tenn.shp", symbol:=FloodFill, color:= moRed
```

Here, the := symbol makes calling the function possible with the arguments in a different order than they are listed in the function definition.

Control Structures and Message Boxes

The sections that follow discuss various control structures and message boxes. These include *If/Then-Else-End If* statements, *Select Case* statements, loops, *For/Next* statements, breakouts, and various message boxes.

If/Then-Else-End If

If/Then-Else-End If statements allow programs to branch. Every *If* must have an *End If*, unless it is all on one line, as in the following.

```
If time > 20.5 then go_home
```

Consider the following example from the web surfing program in Chapter 2.

```
If txtLocation <> "" Then
 WebBrowser1.Navigate txtLocation
Else
 txtLocation.SetFocus
 Beep
End If
```

In this example, if there were a web address in the proper text box, the web browser would attempt to navigate to that location.

(Notice the call to the web browser object's *Navigate* function, which is passed the argument *txtLocation*.) If, however, the *txtLocation* did not contain a string, the *Else* portion of the command would be executed. Note that the *If-Else* control structure must end with an *End If*.

Select Case Statements

You can construct very complex *If/Then-Else-End If* structures, as in the following.

```
If condition1 then
  Outcome1
Else
  If condition2 then
  Outcome2
 Else
   If condition3 then
   Outcome3
Else
   Defaultoutcome
  End If
 End If
End If
```

Although this will work, a simpler approach would be to use a *Select Case* statement. The following shows the structure of such a statement.

```
Select Case condition
 Case condition1
  Outcome1
 Case condition2
  Outcome2
 Case condition3
  Outcome3
 Case Else
  Defaultoutcome
End Select
```

VB evaluates the condition, and then jumps to the appropriate case. The *Case Else* statement is optional. You may not have a *Case Else* in your *Select* statement. *Select Case* options are much easier to understand, and you will be less prone to making logic errors than if you try to enter a series of nested *If/Then-Else* statements.

NOTE: *C programmers might think that there are "break" statements missing in the previous code. However, VB is different from C. When a* Case *statement is found, all statements listed under it are executed until the next* Case *or* End Select *statement is reached. In C, all statements under a case, including those for other cases, are executed until a break statement is encountered.*

Suppose you had a GIS program that allowed the user to edit map layers. Assume you wish to set the caption in a text box to reflect the type of feature being edited. In MO, a map layer has a *shapeType* property. You will work with three layer types: point, line, and polygon. Thus, the code needed to implement this functionality would be as follows.

```
Select Case mapLayer.shapeType
      Case moPoint:
           text1.text = "Point Feature"
      Case moLine
           text1.text = "Line Feature"
      Case moPolygon
           text1.text = "Polygon Feature"
End Select
```

Doing the same thing with *If* statements would require nesting an *If/Then-Else* statement within an *If/Then-Else* statement. As the number of possible cases gets large, the nesting of *If* statements can become quite complex.

"Do" Loops

The first loop to consider is *Do While...* loop. In this type of loop, a condition is tested to see if it is true. If it is, the loop is executed one time. Once the loop is executed, the condition is reevaluated to see if it is true. If it is, the loop is executed again, and so on. Obviously, you must include some statements that can cause the condition to become false or that can break out of the loop. If you do not, the loop will iterate indefinitely.

Chances are you will write a program some day that appears to be doing nothing. It just seems to be stuck. It may be that the condition in a *Do While* loop is never false, whereby the program will happily execute the loop (called an infinite loop) indefinitely (or until you kill the program.) The following code searches a string named *fullFile*, which contains a path and a file name. Here, the objective is to find the name of the file, not including the path, and the location of the period (for the file extension), if there is one. As you will see in Chapter 5, this is useful when searching the string returned from an Open File dialog for the name and location of a shape file.

```
Test = False
textPos = Len(fullFile)
if textPos > 0 then
 Do While Test = False
  textPos = textPos - 1
  tempChar = Mid$(fullFile, textPos, 1)
  If tempChar = "." Then
   periodPos = textPos
  ElseIf tempChar = "\" Or textPos = 0 Then
   Test = True
  End If
 Loop
End If
```

This code starts by setting *Test* (a Boolean variable) to *False* and *txtPos* to the length of the *fullFile* string. This loop continues until it encounters a backslash, indicating the last part of the path, or until it encounters the beginning of the string, indicating that there is no path. The loop also stores the location of the period so that the file extension can be located in the string.

A second type of loop is *Do...Loop While.* This is similar to *Do While...Loop* except that it places the condition statement at the end of the loop. This ensures that the loop is always executed at least one time, whether the condition is true or false. Put another way, *Do While...Loop* tests the condition *before* executing the statements within the loop, whereas *Do...Loop While* tests the condition *after* executing the statements within the loop.

For...Next Loops

For...Next loops use a counter to perform a set of tasks a fixed number of times. The structure is as follows.

```
For counter = start to end [step]
      Stuff to do
Next counter
```

The loop starts with the counter equal to the start value. It processes the commands (*Stuff to do*), and then increments the counter to the next value by a value of *step* (the default is 1). After incrementing the counter, it checks to see if the *counter* value is greater than *end*. If it is, the loop is finished. If not, *Stuff to do* is executed again. The following is an example from an MO program you will write.

```
For j = 0 To Form1.Map1.Layers.Count - 1
  lstLayers.AddItem Form1.Map1.Layers(j).Name
  lstLayers.Selected(j) = Form1.Map1.Layers(j).Visible
Next j
```

This code fragment is used to load the names of all layers in a map into a list of layers. Whether a layer is selected or not (in the list) depends on whether it is visible or not. Let's walk through this code. In this example, a counter, *j*, is set to 0. The first layer is added to the list, and its selected status is set equal to its visibility status. Then the next layer is added, and so on, until all layers are added to the list. The following are a few rules to keep in mind about *For...Next* loops.

- If start = end, the loop runs 1 time.
- If start > end, the loop never runs.
- Listing the counter symbol after the word *Next* is optional. (However, this may help you to know which *For* belongs with which *Next*.)

The *For Each [element] In [group]* loop is similar to the *For Next* structure, except that it executes a statement for each element in a collection or array. The following is another example from a program you will work on later. You have probably used an "identify" function in ArcView or some other GIS. When the user clicks on a feature, a table appears that lists the fields in a database for the current map layer, as well as the values of those fields. The following code shows how to construct the content of such a list.

In this code fragment there is a selected feature stored in the variable *gSelection*. For this feature, you obtain the collection of fields in its attribute database (*gSelection.Fields*). If the field is of type *string*, the name of the field and its value are added to a list (*.fieldlist*), which is displayed in a dialog box. If it is not a string, the name of the field and its value (converted to a string) are added.

This is done for all fields. The following code builds the list of fields and values.

```
For Each curfield In gSelection.Fields
 If curfield.Type = moString Then
   .fieldList.AddItem curfield.Name + " = " + curfield.Value
 Else
   .fieldList.AddItem curfield.Name + " = " + _ curfield.ValueAsString
 End If
Next curfield
```

In Chapter 7 you will develop an identify function that uses this code fragment.

Breaking Out

There are times when you want to break out of a *For* or *Do* loop. To break out of a *For* loop, use *Exit For.* To break out of a *Do* loop, use *Exit Do.*

Message Boxes

Message boxes do just what you would expect: they place messages on the screen. The syntax for using a message box is as follows.

```
MsgBox(prompt[, buttons] [, title] [, helpfile, context])
```

Here, *prompt* is the message you wish to display. The remaining arguments are optional. *Buttons* represents the type of buttons that can be on a message box, such as Yes, No, and Cancel buttons. The various button values are listed in the VB Help file. *Title* places a title on the message box.

You will find it helpful to use message boxes when trying to trace your program. Every programmer at some time needs to see if his program is behaving as it should. Message boxes that display variable values or messages to the programmer are useful ways of accomplishing this.

Summary

In this chapter you have looked at several aspects of VB. Much of this material may seem abstract. If you are new to programming, you may feel that you do not fully comprehend all that is presented. That's understandable. As you work through the remaining chapters of this text, all of which deal with developing mapping applications, the items discussed in this chapter will be used repeatedly. Eventually, much of this material will seem quite familiar.

Chapter 4

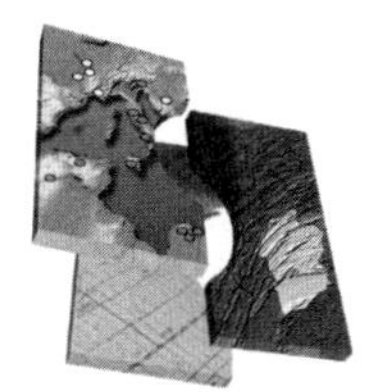

The MapObjects ActiveX Control

Introduction

In this chapter you will begin to work with MapObjects. MapObjects (MO) is an ActiveX control that has a wealth of GIS functionality built into it. By combining this functionality with VB's user interface tools, you will begin to develop a custom GIS system. In the remaining chapters you will add more functionality to the GIS software you start developing in this chapter.

ActiveX Components and DLLs

In VB, subs and functions are pieces of code that perform a certain task or tasks. Suppose you had a task that had to be used in several programs. You could place the code that performs that task into every program you write. However, that would be inefficient. A better alternative is some form of executable code that can be called from each of your programs. This reduces the amount of space taken up by your programs, as several programs can share the same "special executable" code.

This is what DLLs and ActiveX components do. You saw some of these in Chapter 2 in regard to development of the bookstore and web browser applications. The ActiveX control (*dbgrid*) and DLL (*shodocvw.dll*) used in Chapter 2 are not standalone programs. However, they are compiled code. Other programs can use them as needed. If you look in the Windows System32 subdirectory, for example, you will find all sorts of DLLs and OCXs (ActiveX plug-ins). (See figure 4-1.)

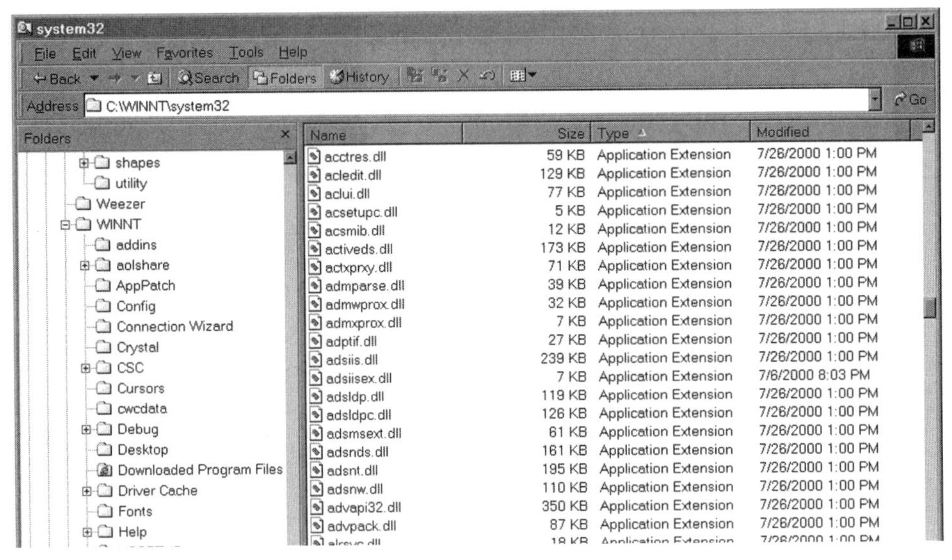

Fig. 4-1. Windows System32 subdirectory.

These components avoid a lot of unnecessary work. It would be inefficient if every Windows program had to have its own code for resizing, minimizing, maximizing, and closing windows.

A DLL is a dynamic link library. A DLL tells any program that calls it: "Here are my functions. If you want to use them, send me the following arguments." A DLL file can contain one function or several functions. Later in this text you will explore alternative methods of serving maps produced by VB/MO programs on the Internet. In Chapter 14, you will work with a method that requires the use of a DLL. Called *jpgegdll.dll*, it can transform a bitmap file to a *jpg* file. This is necessary because web pages can display *jpg* files but not bitmaps. (In Chapter 15, you will use a different method.) The details of this particular DLL are reserved for Chapter 14. The point here is that a DLL exposes functions to other programs.

ActiveX components (also called OCXs, or components) are a step up the software evolutionary chain from DLLs. ActiveX components, like MO, expose *objects* to a calling program and create a *client/server relationship* between your code and the OCX. Because they expose objects, the calling program can access *methods*, *properties*, *constants*, and *enumerations* of the objects.

The ActiveX control acts like a server (called an OLE automation server), taking requests from a client (your program) and returning the desired result. There is a type of DLL, called an extension

DLL, that does expose classes. You used just such a DLL in Chapter 2 when you constructed the web browser. However, DLLs do not have the client/server relationship with your program that ActiveX components do.

The MapObjects ActiveX Control

***CD-ROM NOTE**: The VB project in the* Chapter4_1 *directory on the companion CD-ROM starts here.*

In this section you will explore the MO ActiveX control. Toward this end, try the following.

1. Start a new VB project.
2. Right click on the toolbox to add a new component, and add an ESRI MapObjects 2 component (see figure 4-2).
3. Click on the MapObjects icon in the toolbox, and then press F1 (see figure 4-3).

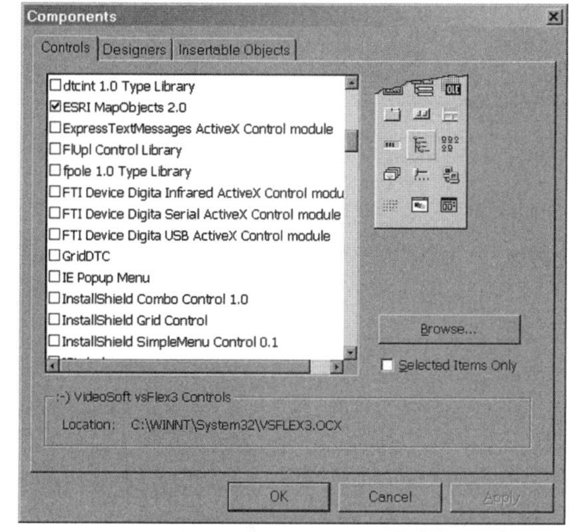

Fig. 4-2. MapObjects 2 component.

This will open the MapObjects Help file.

***NOTE**: The help file for MapObjects is not part of the VB Help file.*

4. Place an MO map control form on your main program form (form *Form1*). To do this, make sure the MO control icon is selected in the toolbar and draw a rectangle on form *Form1*.
5. Right click on the map control and select Properties. In the General tab, select Add Layers.

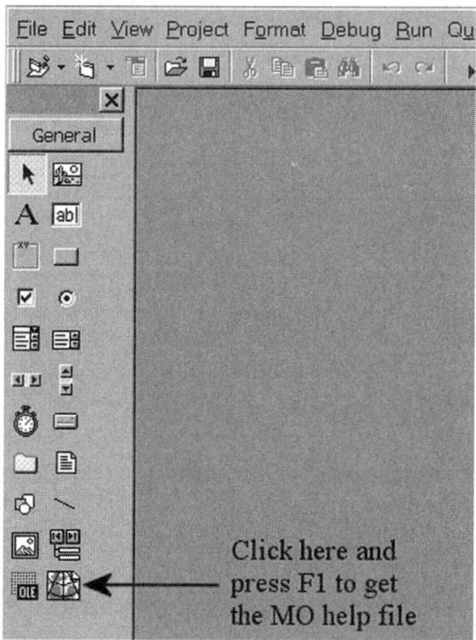

Fig. 4-3. MapObjects Help file.

This text comes with some layers for the United States.

6 Navigate to where those layers are located (either on the companion CD-ROM in the *shapes/USA* directory or on your hard disk if you copied them from the companion CD-ROM), and select the layers *States, Roads,* and *Lakes.*

7 Double click on a layer to bring up its properties box (see figure 4-4).

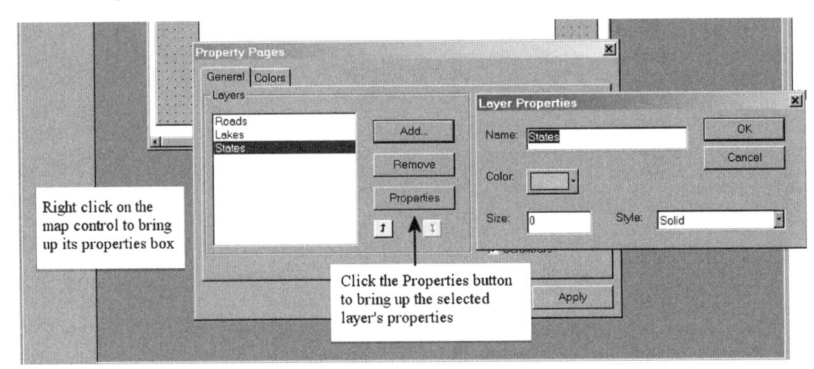

Fig. 4-4. Layer Properties box.

8 Set the colors to whatever you want. Set the *States* layer at the bottom of the list, the *Lakes* layer next, and the *Roads* layer at the top.

9 Run the program.

Your program should produce the map shown in figure 4-5.

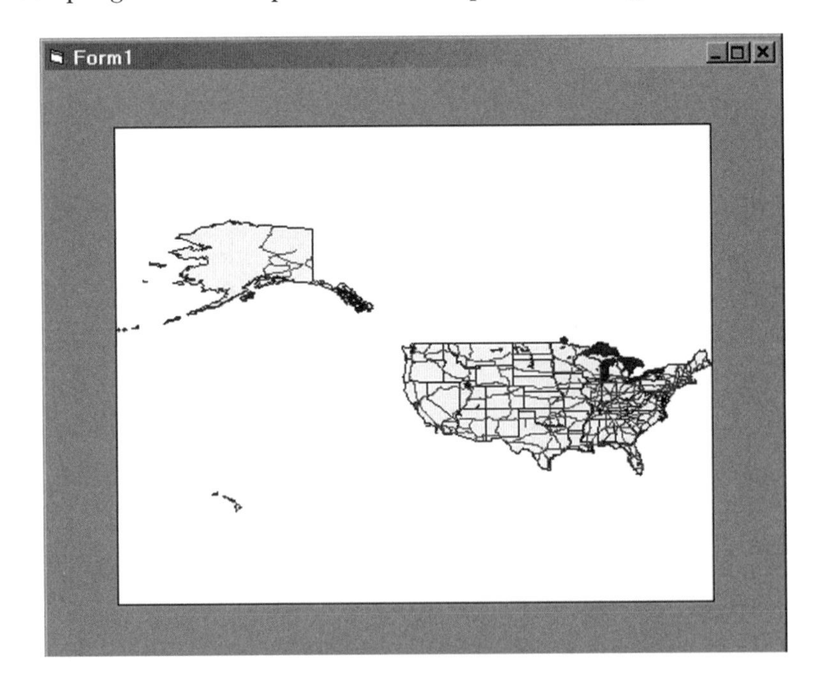

Fig. 4-5. Map as a result of running the program.

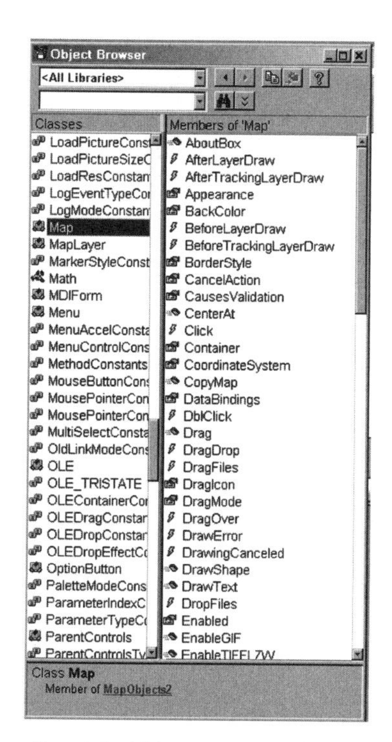

Fig. 4-6. Object Browser for the Map control.

The map control's (the box you put on the form) property box gave you access to some properties: the layers to add, their drawing properties, and the color palette to use. However, these are not everything that a map control exposes to your program.

10 Close your program.

11 Press the F2 key to open the object browser.

12 Set the object type to Map Objects, and click on the Map class (left-hand panel).

This accesses a list of objects and their properties, methods, classes, constants, and enumerations the Map class exposes to VB (or any other Windows-compliant programming language). As you can see in figure 4-6, there are a lot of things to learn!

MO has five major types of objects: data access objects, map display objects, address-matching objects, projection objects, and geometric objects. Each of these types of objects consists of several objects, and each of those objects has properties and methods. Data access objects are used to open and add layers and to work with database tables. Map display objects are used to control the display of map layers. Address-matching objects are used for geocoding. Maps can be projected, and map layers with differing projections can be combined with projection objects.

Geometric objects consist of different types of entities, such as lines, polygons, rectangles, and points. These objects are combined to add functionality to your maps. The clearer you are about how these object types are used and how they interact, the easier it will be for you to develop your own applications. The purpose of the remaining chapters is to familiarize you with many of these objects and their properties and methods.

When you display layer information on a map, you work with (at a minimum) the Map Control, the Layers Collection, and at least one of the following types of data objects: a *MapLayer* object (vector data), an *ImageLayer* object (raster data), and/or a *TrackingLayer* object (dynamic event data). Each of these objects has methods and properties of its own, and can contain other objects (much like a *form* object can contain a *listbox* object.) For example, you will see that a *layer* object can have a *renderer* object.

If you go to the directory *C:\Program Files\ESRI\MapObjects\Samples\Diagram*, you will find a diagram of all object relationships in MO. (If you cannot find this diagram on your computer, you can download a copy from *http://www.esri.com/devsupport/mapobjects/files/other/modiagram.exe.)* There is a method for reading this diagram. To the left of each object are its methods. To the right are its properties. A double-headed arrow between an object and its methods indicates that the method returns a value. A single-headed arrow pointing to the object indicates that the method does not return a value.

For properties, a circle on the end of each line indicates that the property is read/write (i.e., you can get its value and set its value). A circle only on the left (near the object) means it is a write-only property, whereas a circle on the right means it is a read-only property. If you look at the diagram that comes with MO, you will see that for the map control, the method *FromMapDistance* returns a value, whereas the method *FlashShape* does not. You also should see that *Layers* is a read-only property, and that *BackColor* is read/write. (See figure 4-7.)

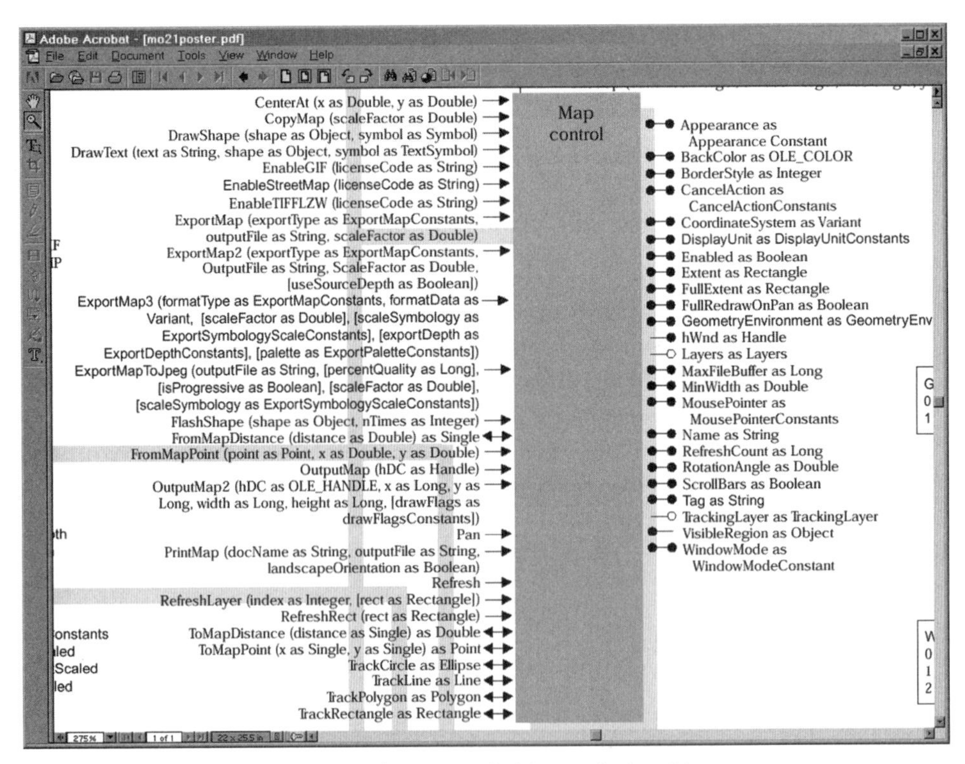

Fig. 4-7. MapObjects diagram of object relationships.

Each of these objects has a set of properties and methods. Let's take a closer look at the Map Control object. It is unlikely that you will use all events, methods, and properties associated with this object. However, you will use many of them. The following are properties, methods, and events you should know.

- *ExportMap:* Copies the current map to the Windows clipboard.
- *MouseDown:* Occurs when the user clicks on the map. It records which mouse button was pressed and the location of the mouse on the map.
- *Pan:* Allows you to pan the map.
- *FlashShape:* Makes a shape flash on the screen. You can set the number of times the shape should flash.
- *TrackRectangle:* Allows you to rubber-band a rectangle on the map.
- *Extent:* Current extent of the map.
- *FullExtent:* Bounding box for all layers on the map.
- *Refresh:* Forces a full redraw of the map.

When you create a map window on the form (draw a map object in VB), you create two other types of objects automatically. The first is a Layers collection; the second is a tracking layer object. Each map control has one and only one Layers collection. The collection is at first empty, but will eventually be populated with your map layers.

When working with layers, the *Refresh* method does not always have to be invoked. Some actions automatically fire the *Refresh* command. These are:

- When a layer is added to the Layers collection
- When the *Pan* and *CenterAt* methods are used
- When scroll bars are invoked
- When the map extent changes
- When the *Clear* or *Remove* methods are used on the Layers collection

Now that you know a bit about the Map Control, let's look at the Layers collection. The Layers collection is a VB collection that contains the name of all vector and image layers. The events, methods, and properties associated with this collection are fairly obvious. You add layers to the collection with the *Add* function, and you remove them with the *Remove* function. A layer can be accessed via an index value, and (this is one of the strengths of a collection) the index value can be an integer or a key value, which for layers is the layer name. The *Move* functions are used to change the order in which layers are drawn.

There is a default method for the Layers collection: *Item.* That is, if no method is listed, the map object assumes you want to use the *Item* method. Thus, the following are equivalent statements. In this example, *Lakes* has an index value of 2.

```
Map1.Layers.Item(2).Symbol.Color = vbBlue
Map1.Layers(2).Symbol.Color = vbBlue
Map1.Layers.Item("Lakes").Symbol.Color = vbBlue
Map1.Layers("Lakes").Symbol.Color = vbBlue
```

You now know a bit about layers as a collection, but what about an individual layer? When you select an item from the collection, you get either a *MapLayer* object (vector) or an *ImageLayer* object (raster). Thus, *Map1.Layers(2)* in the preceding code refers to a *MapLayer* object. The following are useful methods of the *MapLayer* object.

- *Add Relate:* Allows relational joins between a layer's attribute file and a database file.
- *SeachExpression:* Supports queries of a layer's attribute file.
- *SearchByDistance:* Supports spatial queries based on distance from an object.
- *SearchShape:* Supports spatial queries based on the relationship between map features. Such relationships would include touch, intersect, and containing.

The following are properties of the *MapLayer* object.

- *Extent:* Bounding box of the layer.
- *LayerType:* Layer type; either an image layer or a map layer.
- *Name:* Name of the layer.
- *ShapeType:* Type of map layer, such as point, line, or polygon.
- *Visible:* True if the layer is drawn on the map; false otherwise.

You will work more with these and other properties as examples become more complex as you progress through the book. The other type of layer you can have in a Layers collection is an *ImageLayer* object. You might want to study its object diagram, methods, and properties. (The examples in the tutorials of this book do not use image layers.) By the time you have finished this book, working with image layers will seem straightforward.

▪▪ Working with Layers

In this section, you will build an application similar to the previous. However, here you will add a button that will allow the user to turn layers off or on.

CD-ROM NOTE: *The project in the* Chapter4_2 *directory on the companion CD-ROM starts here.*

1. Start by creating a project similar to the previous.

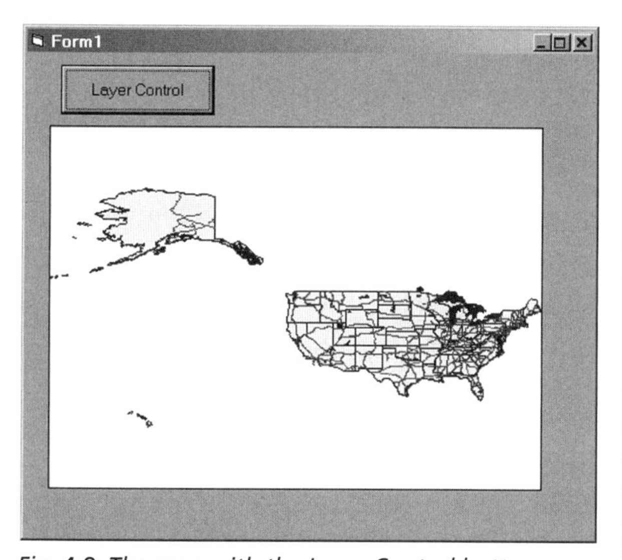

Fig. 4-8. The map with the Layer Control button.

Be careful in creating new projects. If you simply click on File > Save Project As in the VB development environment, here is what will happen. New project and workspace files will be created. However, they will refer to the forms you have in your original project. If you change the forms in the new project, your original project will *not* behave as it once did. In fact, it may not work at all. If you want to keep a clean copy of your original project, exit VB, and copy the content of the current project to a new directory. Work on that new copy.

2. Once you have a new project, add the following layers: *Cities*, *Lakes*, *Roads*, and *States*. Be sure that the *Cities* layer is at the top of the list of added layers, and that *States* is at the bottom.
3. On the main form (the one with the map object), add a command button. Label the button *Layer Control*.

Your program should look like that shown in figure 4-8.

What do you want the layer control button to do? If you are an experienced GIS user, you can probably think of many things you would like this button to do. However, let's start with a relatively simple task: turning layers off or on. When the user presses the button, a new form should pop up. The new form should have a check box list that contains the name of every layer currently in the map. If the layer is currently visible, the check box should be checked. If it is not visible, the check box should not be checked. If the user checks a layer off, the layer should be made invisible. If the user checks a layer on, the layer should be made visible.

4. Create a new form. To do this, click on Project > Add Form, and then choose a standard form. The form will automatically be named *Form2*.
5. To this new form, add a list box control. Set the list box name to *lstLayers*, and the Style to *1-Checkbox*.

Now let's add some code to display the new form when the user clicks on the Layer Control button.

6. Go to form *Form1*'s code page and add the following code:

```
Private Sub cmdLayers_Click()
 Form2.Show
End Sub
```

Whenever the user clicks on the Layer Control button, form *Form2* will pop up.

7. Go to form *Form2*'s code page and add the following code:

```
Dim FormUp as Boolean
Private Sub Form_Load()
 Dim i As Integer
 FormUp = True
 For i = 0 To Form1.Map1.Layers.Count - 1
  lstLayers.AddItem Form1.Map1.Layers(i).Name
  lstLayers.Selected(i) = Form1.Map1.Layers(i).Visible
 Next i
 FormUp = False
End Sub
```

This function, *Form_Load*, is called before the form is put on the screen (it loads it in memory). You use this function to make sure the form and the map are synchronized before the form is displayed. The *FormUp* variable is a Boolean (true/false) that checks to see if the form is being loaded or if it is already on the screen.

The *Form_Load* function contains a loop over each layer in the map. You start by getting each layer's name. To get the name, you must know the layer; to know the layer, you must know the map control; and to know the map control, you must know the form. Hence, you get the name with the following statement.

```
Form1.Map1.Layers(i).Name
```

Put another way, *Name* is a property of the layer *Layers(i)*, the *i*th element of a collection of layers. The collection of layers is a property of the *Map1* control. The *Map1* control is an object belonging to *Form1*. Notice how this is interpreted from right to left. As you gain more programming experience, you will find this right-to-left reading aids in interpreting statements that have many objects in a single statement.

Once you have the name, you add it to the list of layers (*lstLayers.AddItem*). Next, you check the current layer's visible property, which will be either true or false. You set the check box status to that value (either true or false).

Now that you have populated the layers, you can turn your attention to what happens when a layer is clicked on or off. The following is the code that should be added to form *Form2*'s code page.

```
Private Sub lstLayers_ItemCheck(i As Integer)
 If Not FormUp Then
  Form1.Map1.Layers(i).Visible = lstLayers.Selected(i)
  Form1.Map1.Refresh
 End If
End Sub
```

If *FormUp* is false (you have already put the form up), you simply toggle the visibility status of a layer to its selected status. You then need to refresh the map to make sure the new status is reflected on the map. To reinforce some of these concepts, you might want to try adjunct exercise 4-1, which follows.

Adjunct Exercise 4-1: Working with Layer Functions

To make sure you understand what these functions do, try the following.

1 Comment out the *If Not FormUp* and *End If* lines in *lstLayers_ItemCheck*.

2 Run the program.

As the form loads, it adds items to the list. Adding items to the list causes an *ItemCheck* event to be fired. If you take out the lines in step 1, the map will refresh every time a layer is loaded, causing your map to be drawn as many times as there are layers.

3 Reinstate the lines commented out in step 1 and comment out the *Form1.Map1.Refresh* line.

When you run the program this time, you can check layers on or off, but the map never changes. There must be an explicit call to the *Refresh* method in order to have the map reflect your visibility settings.

4 Keep the *Form1.Map1.Refresh* line commented out in the *lstLayers_ItemCheck* sub. Add the following sub to form *Form2*'s code page.

```
Private Sub Form_Unload(Cancel As Integer)
 Form1.Map1.Refresh
End Sub
```

When you run the program this time, the map will reflect your visibility changes, but only after you close form *Form2*. Thus, if you turn off the *Roads* and *Lakes* layers, that change will not be reflected until you close form *Form2*.

Summary

In this chapter you learned how to add a map control to a VB form. You used the properties box associated with that control to add layers to the map and to set the layer drawing colors and order. Properties, methods, and events associated with the map control, the Layers collection, and the *layer* object were presented.

In the second program in this chapter, you created a form that will come up in response to a Layer Control command button. The *Form_Load* function for form *Form2* was constructed so that the list of layers and each layer's visibility status were synchronized between the map and form *Form2*'s check box list. In Chapter 5 you will continue to develop more functionality for controlling map layers.

Chapter 5

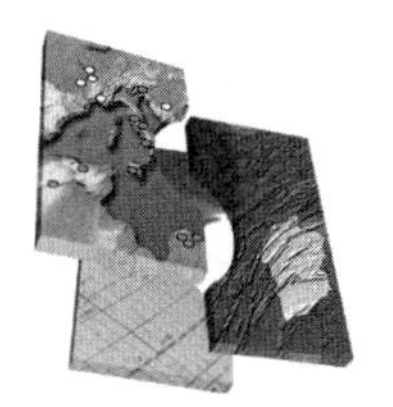

Managing Map Layers

Introduction

Before exploring new topics, let's take a closer look at what is going on in the last example of Chapter 4. You should understand the use and placement of the variable *FormUp*, the use of the argument passed to the *ItemCheck* method, and the method of turning layers on and off. If you completed the adjunct exercise at the end of Chapter 4, you should have a good understanding of these issues, elaborated upon in the sections that follow.

The Use of FormUp

When someone clicks on the Layer button, the form with the layer list appears. Before it appears, the form's *Load* function is called. It is in this function that you populate the list box and set the visibility status. Every time you set the visibility status, you update the check box in the corresponding list item. Every time you update a check box, Visual Basic (VB) fires an *ItemCheck* event. Most of the time you want a change in the status of a check box to cause the map to be refreshed, but not always. In this case you do not want the map redrawn when form *Form2*'s *Load* function is called.

To keep from unintentionally firing the *Refresh* method, in Chapter 4 you defined a Boolean variable (*FormUp*) to track whether or not the form is loading. You want this variable to be seen by two subs (*Form_Load* and *Item_Check*), and therefore you gave it module-level scope. That is, you placed the following line at the top of the file, outside any sub or function.

```
Dim FormUp As Boolean
```

Whenever *Form_Load* is called, *FormUp* is set to True. Just before *Form_Load* exits, *FormUp* is set to False. In *ItemCheck*, you reset the visibility of a layer and redraw the map when *FormUp* is False. Therefore, the first line of this sub was as follows.

```
If Not FormUp Then
```

The ItemCheck Argument

You might have been tempted to use a *For* loop to check the status of the layers whenever an item in the layer list is checked (on or off). Take a close look at the declaration of the *ItemCheck* function:

```
Private Sub lstLayers_ItemCheck(item As Integer)
```

This function gets passed an integer named *item*. This value is the index number of the item whose check status has changed (been turned off or on). There is no need to loop through all layers because you already know which item in the list has been checked on or off. Because the list was built to correspond exactly to the order in which the layers are drawn, you know which layer you should turn off or on.

Method of Turning Layers Off or On

You might be tempted to use statements such as the following to turn layers off or on.

```
If LstLayerList.Selected(j) = True then
   From1.Map1.Layers(j).Visible = True
Else
   From1.Map1.Layers(j).Visible = False
End If
```

This would work, but it is not the best method. Both *From1.Map1.Layers(j).Visible* and *lstLayerList.Selected(j)* are Boolean variables. That is, they take or return a value of True or False. Is there any need for all of these *If/Then-Else-End If* statements? No. Using *Form_Load*, you simply need to write:

```
Form1.Map1.Layers(j).Visible = lstLayers.Selected(j)
```

VB's Common Dialog

Before getting into more MapObjects material, let's look at the common dialog control in VB. Common Dialog is a component that can be added like any other control, by activating (checking) the Microsoft Common Dialog Control option in the Add Components dialog. You place a common dialog control on a form much like any other control. However, when your program runs, you do not see the control. It is hidden. You must have it pop up in response to a command button or some other event.

Common dialogs are very helpful. They contain procedures for many common (hence the name) features you need in programs, such as File Open dialogs, File Save, Color, Font, Print, and Help. Take a look at VB's Help file on common dialogs. You will see that the method used for opening a common dialog will determine the type of dialog opened. Table 5-1, which follows, outlines these dialog types and the methods used to open them.

Table 5-1: Common Dialog Types Opened by Various Methods

Method	Dialog Displayed
ShowOpen	Show Open dialog box
ShowSave	Show Save As dialog box
ShowColor	Show Color dialog box
ShowFont	Show Font dialog box
ShowPrinter	Show Print or Print Options dialog box
ShowHelp	Invokes the Windows Help Engine

You can create a pop-up File Open dialog via code such as the following. This component saves a lot of coding.

```
CommonDialog1.ShowOpen
```

Data Connections, GeoData Sets, and Adding Layers and Images

Let's turn our attention to map objects. To this point, you have added layers to a map by hard-coding them into a program at design time. However, what if you wanted the user to be able to add new layers at run-time? For this you would have to find and add the proper shape files. The same is true for adding images.

It may seem that all you have to do is use the Common Dialog control to select an **.shp* file and add it to the layers collection. However, there is more to it than that. First, a shape "file" is really a set of *at least* three files: an *.shp*, an *.shx*, and a *.dbf* file. Second, the *.dbf* file is a database file, and the *.shp* and *.shx* files have special formats. Fortunately, MO takes care of a lot of the details for you. However, there is some work you must do.

There are three new objects you need to use: the *DataConnection* object, the *GeoDatasets* collection, and the *GeoDataset* object. This may seem like a lot of overhead for opening a few files, and it is. However, MO supports several spatial data formats (shapes, CAD drawings, images, and coverages), which can be read from a disk or accessed via ESRI's Spatial Database Engine (SDE).

Assume for the purposes of this book that the *DataConnection* object connects programs to a disk drive and folder. Within that folder there may be many shape files (the *GeoDatasets* collection), of which you would choose one (the *GeoDataset* object). Let's take a closer look at these objects.

The DataConnection Object

The *DataConnection* object represents either a folder on a disk or a connection to a server. You create a data connection via the following code.

```
Dim dataconn As New MapObjects2.DataConnection
dataconn = "c:\esri\esridata" 'this is just an example
```

The first line declares a *dataconn* as a *DataConnection* object. *New* is used because you are creating a new object. Notice, however, that for variables you do not have to use the keyword *New*. The program automatically understands that a variable takes up a certain amount of memory space. VB simply creates a new variable and allocates space. However, creating objects involves more work. The object needs space for its properties and methods. Further, the properties and methods must be associated with each instance of the object. VB must construct the object, allocating the space it needs. The term *New* is needed for VB to accomplish these tasks.

The second line sets the path to the location of the shape files. To understand the properties of the data connection object, see the object diagram that comes with MapObjects (see figure 5-1).

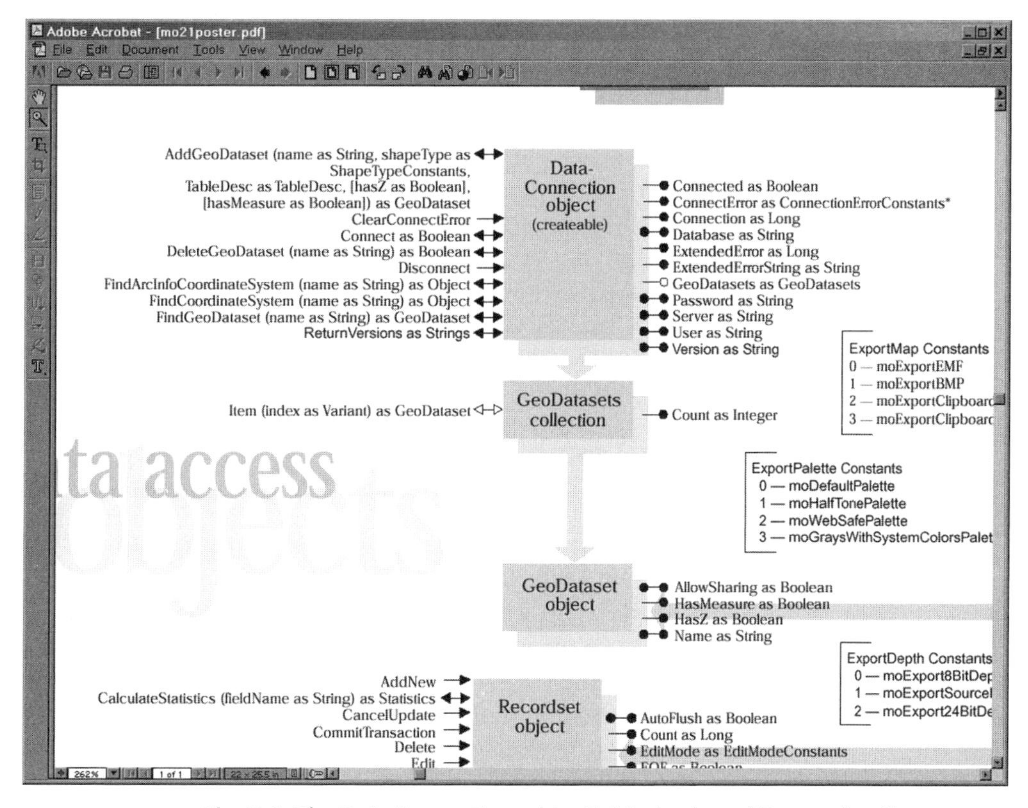

Fig. 5-1. The DataConnection *object's Methods and Properties diagram.*

The GeoDataset Object and Collections

Once you have a path to a folder containing shape files, you need to create a *GeoDataset* object (the shape file group), such as follows.

```
Dim gSet As GeoDataset
dataconn.Database = "c:\esri\esridata"
'Set Database property of DataConnection
If dataconn.Connect Then
  Set gSet = dataconn.FindGeoDataset("Lakes")
```

This code contains a *DataConnection* object. If you can connect to it (i.e., if you can read the directory or connect with the server), you use the *DataConnection* object's procedure *FindGeoDataset* to return a named *GeoDataset* (in this case, *Lakes*), and set the *GeoDataset* (*gSet*) equal to that result. That is, the last line will connect the program to a shape or coverage named *Lakes*.

Using the CommonDialog Object with GeoData Sets

Can you see how you would use the *CommonDialog* object to get layers into your program? Let's see what would have to be changed in the previous code to make it more general. That is, what parts of the code fragment would have to be changed to allow the user to choose any shape file on a disk?

In the previous code, you would want to replace *"c:\esri\esridata"* with a path the user could specify, and *("Lakes")* with the name of a shape or coverage. Where can you get a path and a file name? From the *CommonDialog* object's *FileOpen* box! The procedure for doing this follows.

- ❒ On form *Form2*, create a button that pops up a File Open dialog. Make sure the dialog is set to open the proper type of file (you will see how to do this later). Assume you are looking for shape files. Let the user use Windows tools to select such a file. Assuming the user selects a shape file, the *FileOpen* dialog object will return a string such the following: *"c:\esri\esridata\usa\roads.shp"*.
- ❒ Parse the string returned by *FileOpen* into its path (from the beginning up to the last backslash), its file name (from the last backslash to the period), and its file type (after the period).
- ❒ Use the path to create the data connection, and the file name in the *FindGeoDataset* function.
- ❒ Add the *GeoDataset* to the *MapLayers* collection.

Adding image files is much easier. An *ImageLayer* object has a *File* property, which lists the file name (including the path) of the image file. There is no parsing to be done. Further, because images do not have databases connected to them, there is no need for a *DataConnection* object. The following code, taken from ESRI's MapObjects *MoView2* sample project, demonstrates how you would add image files to a project.

```
Dim iLayer As New ImageLayer
 CommonDialog1.Filter = "Windows Bitmap (*.bmp)| _
    *.bmp|TIFF Image(*.tif)|*.tif"
 CommonDialog1.FilterIndex = 1
 CommonDialog1.CancelError = True
 On Error Go to ErrHandler
```

```
 CommonDialog1.ShowOpen
If CommonDialog1.filename <> "" Then
  iLayer.File = CommonDialog1.filename
  ' move the existing layer to the top
  If Map1.Layers.Add(iLayer) Then
    Map1.Layers.MoveToTop 1
  End If
 End If
 Err Handler:
  Exit Sub
```

In this code, the *CancelError* property is set to True. This allows the program to jump to another part of the sub (*ErrHandler*) when the user clicks on the Cancel button in the Open File dialog. If the *CancelError* property were not present, and the user clicked on Cancel, a layer the user did not want might be added. You will encounter this approach again when adding shapes. If you plan to work with image layers, you should check out the image layer object section of the MapObjects object diagram.

MoView2: A Helpful Reference

MapObjects comes with many sample projects for different languages (e.g., VB, C++, Delphi, and PowerBuilder). This section explores the VB project *MoView2*. You might want to copy the folder *C:\Program Files\ESRI\MapObjects2\Samples\VB\MoView2* to your own folder. This will keep you from changing the original version of *MoView2*. If you cannot find *MoView2* on your computer, you can download the sample programs (and sample data) from ESRI's web site at *http://www.esri.com/download/mapobjects/mo2samples.exe.*

In the following you will create new projects, copy and paste parts from *MoView2* for use in the projects, and then edit those parts. The easiest way to copy and paste from *MoView2* to your own programs is to open two sessions of VB. One session will have *MoView2*, the other your program.

Adding Layers Interactively

CD-ROM NOTE: *The VB project in the* Chapter5_1 *directory on the companion CD-ROM starts here.*

In this section you will use some code in *MoView2* to update the program (the one with the list of layers) you created at the end of Chapter 4.

1 Open your project.

By now you should be able to display the map layers and click them on and off.

2 Add a Common Dialog control, named *CommonDialog1*, to form *Form2*.

3 Add a command button to form *Form2*, and name it *cmdAdd*.

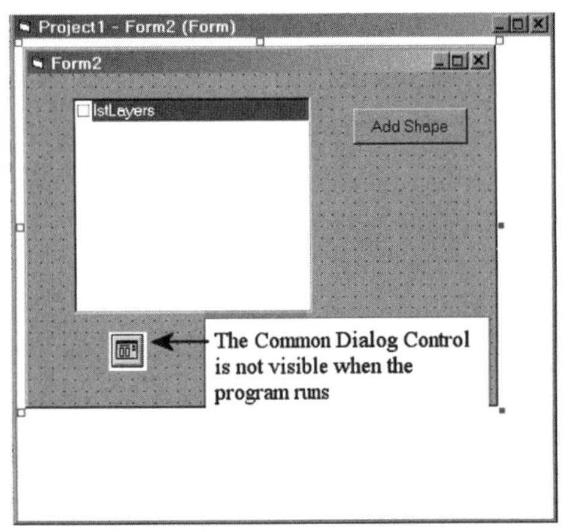

Fig. 5-2. Common Dialog control and Add Shape button added to form Form2.

Your form should now look like that shown in figure 5-2.

4 This new button will be used to add layers to the map; therefore, give it the caption *Add Shape.*

5 In the code for the layer control form, add the following subroutines.

NOTE: *This edited code is based on* MoView2. *In particular, the functions listed here are only for loading shapes. (The sections for coverages and images have been commented out.) This code is discussed in material to follow.*

```
Private Sub cmdAdd_Click()
 addFile
End Sub
Private Sub addFile()
 'For processing into the Layers collection.
Dim fullFile As String, path As String, tempChar As String, ext As _
   String
 Dim Test As Boolean
 Dim textPos As Long, periodPos As Long
 Dim curPath As String
 'Execute common dialog for selecting a file to open.
 Dim strShape As String
strShape = "Shape files (*.shp)|*.shp"
CommonDialog1.Filter = strShape
 CommonDialog1.DialogTitle = "Select file for new layer"
 CommonDialog1.ShowOpen
```

```
 'We have the full path name from the common dialog.
 'Parse out base path.
 If CommonDialog1.filename = "" Then Exit Sub
 fullFile = Trim$(CommonDialog1.filename)
  textPos = Len(fullFile)
 Test = False
 'This loop goes backward through the string
 'searching for the last back slash.  This marks
 'the base path from the returned string.
 Do While Test = False
  textPos = textPos - 1
  tempChar = Mid$(fullFile, textPos, 1)
  If tempChar = "." Then
   periodPos = textPos
  ElseIf tempChar = "\" Or textPos = 0 Then
   Test = True
  End If
 Loop
  'Path is the part of the full file string up to the last backslash.
 curPath = Left$(fullFile, textPos - 1)
 'Send the file name to the procedures that add the layers...
 Dim filename As String
 filename = CommonDialog1.FileTitle
  'Check for file extension. If extension is *.shp, assumed to be
  'shape file. Otherwise, it will be processed and checked as
  'an image file.
 ext = LCase(Mid$(fullFile, periodPos + 1, 3))
  If ext = "shp" Then
 Call addShapeFile(curPath, filename) 'Add shapefile into
 'Layers collection
 'The MoView2 sections for coverages and images are
 'commented out
 'ElseIf ext = "adf" Or ext = "pat" Or ext = "rat" Or ext = "tat" Then
 'Call addCoverageTable(curPath, filename)
 'Else
 ' Call addImageFile(fullFile)
'Add image file into Layers collection
 End If
End Sub
Private Sub addShapeFile(basepath As String, shpfile As String)
 'This procedure validates and adds a shape file to the
 'Layers collection.
 Dim dCon As New DataConnection
 Dim gSet As GeoDataset
dCon.Database = basepath    'Set Database property of DataConnection
 If dCon.Connect Then
```

```
  shpfile = GetFirstToken(shpfile, ".")
  'Extracts suffix of shpfile string
  Set gSet = dCon.FindGeoDataset(shpfile)
  'Finds shapefile as GeoDataset in DataConnection
  If gSet Is Nothing Then
    MsgBox "Error opening shapefile " & shpfile
   Exit Sub
  Else
   Dim newLayer As New MapLayer
   newLayer.GeoDataset = gSet
   'Sets GeoDataset property of new MapLayer
   newLayer.Name = shpfile
'Sets Name property of new MapLayer
   Form1.Map1.Layers.Add newLayer
   'Adds MapLayer to Layers collection
   End If
  Else
   MsgBox ConnectErrorMsg(dCon.ConnectError), vbCritical, _
     "Connection error"
  End If
End Sub
```

When the user clicks on the Add button, the sub *addFile* is called. Here is where all the action is. The *addFile* sub begins by declaring a set of string variables. Their meaning will become clearer in material to follow. The next part of the sub to consider is the following.

```
Dim strShape As String,
 CommonDialog1.Filter = strShape
CommonDialog1.DialogTitle = "Select file for new layer"
 CommonDialog1.ShowOpen
```

The previous code creates a variable to hold the search parameters for shape files. (The code in *MoView2* includes search strings for coverages, CAD drawings, and images.) Let's look at the search parameter for shapes, which follows.

```
StrShape = "Shape files (*.shp)|.shp"
```

When passed to the *OpenFile* dialog object, the first part of the string is a prompt [*Shape files (*.shp)*], and the second part is the file filter (used to filter out files that do not have the proper extension). Figure 5-3 illustrates their use.

These strings are used to build a filter for the Open File dialog. Once the user closes the Open File dialog, the program checks to see if they have chosen anything, as follows.

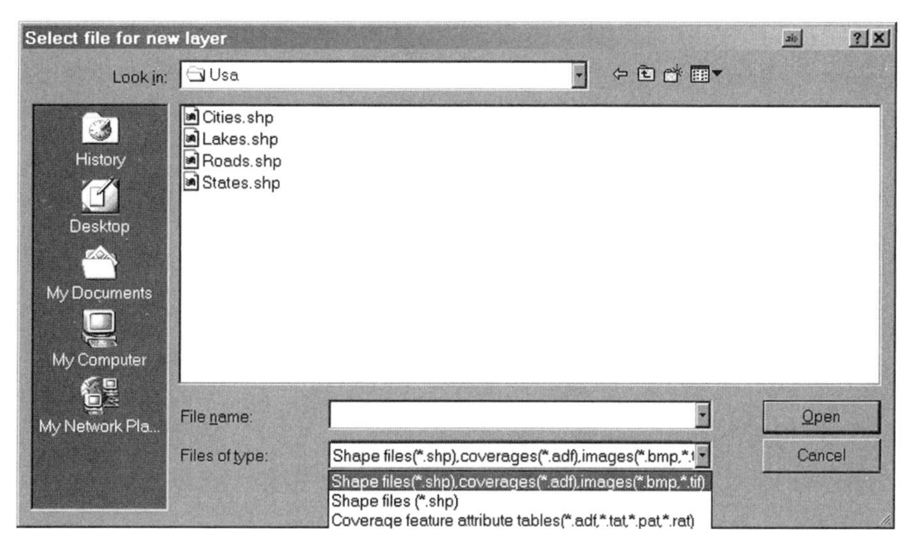

Fig. 5-3. The Open File dialog for returning shapes.

```
'We have the full path name from the common dialog.
'Parse out base path.
 If CommonDialog1.filename = "" Then Exit Sub
 fullFile = Trim$(CommonDialog1.filename)
```

If the user has chosen something, the program loops through the string returned by *OpenFile* (*fullFile*), looking for the period before the extension and the last backslash or beginning of the string. Note that the string is parsed from the end (right to left), and therefore the first backslash encountered is the last backslash in the string.

```
textPos = Len(fullFile)
Test = False
'This loop goes backward through the string,
'searching for the last backslash. This marks the
'base path from the returned string.
Do While Test = False
  textPos = textPos - 1
  tempChar = Mid$(fullFile, textPos, 1)
  If tempChar = "." Then
   periodPos = textPos
  ElseIf tempChar = "\" Or textPos = 0 Then
   Test = True
  End If
 Loop
```

The program notes the location of the period (*periodPos*) and breaks out of the loop once it knows where the backslash is (if there is one). Thus, it knows the locations of the extension, path, and file name. The program then stores the file name, path, and extension. (You might want to review the uses of the string methods *Trim, Left,* and *Mid.*) If the extension is *shp,* the program calls *addShapeFile.* Otherwise, it does nothing. (Subs that would add other types of layers are commented out. To see these subs, examine *MoView2.*)

addShapeFile gets passed the path and the file name. The extension does not need to be sent to *addShapeFile* because the calling sub ensures that the file refers to a shape. In the *addShapeFile* sub, a new *DataConnection* object gets created and is set to the path. A new *geodataset* object also is created, and is set equal to the file name.

```
dCon.Database = basepath
'Sets Database property of DataConnection
 If dCon.Connect Then
  shpfile = GetFirstToken(shpfile, ".")
'Extracts suffix of shpfile string
  Set gSet = dCon.FindGeoDataset(shpfile)
'Finds shapefile as GeoDataset in DataConnection
```

The program discerns whether or not the shape exists. If the shape does exist, a new layer object is created and its *GeoDataset* property is set to *gSet.* The layer's name is added to the list of layers and the layer is added to the map.

```
Dim newLayer As New MapLayer
newLayer.GeoDataset = gSet
'Sets GeoDataset property of new MapLayer
newLayer.Name = shpfile
'Sets Name property of new MapLayer
Form1.Map1.Layers.Add newLayer
'Adds MapLayer to Layers collection
```

If the program cannot connect to the *geodataset* property, an error message is issued. Note that when a shape is added to a map it becomes by default the topmost layer.

The previous functions referenced some subs and values that are not currently defined. To implement these, you need to create two new modules: *modStringHandler* and *modUtility.* To create these, perform the following.

6. Click on Project, and then on Add Module. Do this two times to create two new code pages.
7. Name these new modules *modStringHandler* and *modUtility.*

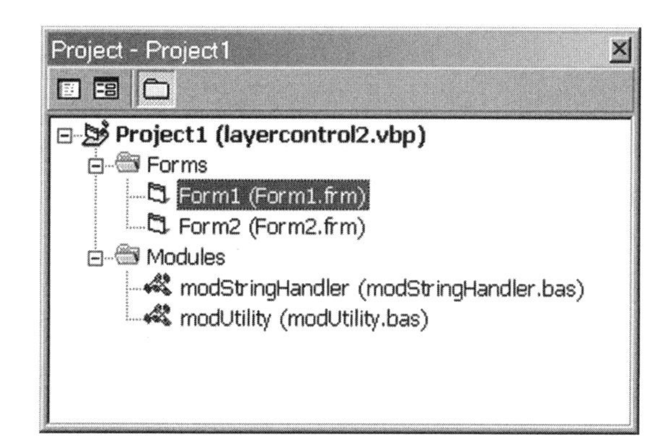

Your Project Explorer window should like that shown in figure 5-4.

8 Add the following code to the *modStringHandler* module.

Fig. 5-4. New modules in the Project Explorer window.

```
Option Explicit
Function GetFirstToken(StrIn As String, Delim As
String) As String
 ' Gets the portion of String "S", up to the
 ' first occurrence of delimiter "D"
 ' Returns String token "T"
 Dim Split As Long
 Dim Tok As String
 StrIn = Trim$(StrIn)
 Split = InStr(1, StrIn, Delim)
 If (Split <= 0) Then
  ' No delimiter in the string.
  ' Return the whole thing.
   Tok = StrIn
 Else
   ' Get everything up to the first delimiter.
   Tok = (Trim$(Left$(StrIn, Split - 1)))
 End If
 GetFirstToken = Tok
End Function
```

9 Add the following to the *modUtility* module.

```
Option Explicit
'ConnectErrorMsg - Defines an appropriate error message for
'the DataConnection object
Public Function ConnectErrorMsg(errNum As Integer) As String
 Select Case errNum
  Case moNoError:   ConnectErrorMsg = "No Error"
  Case moUnknownError:  ConnectErrorMsg = "Unknown Error"
  Case moAccessDenied:  ConnectErrorMsg = "Access Denied"
  Case moInvalidUser:  ConnectErrorMsg = "Invalid User"
  Case moNetworkTimeout: ConnectErrorMsg = "Network Timeout"
  Case moInvalidDatabase: ConnectErrorMsg = "Invalid Database"
  Case moTasksExceeded: ConnectErrorMsg = "Tasks Exceeded"
```

```
  Case moFileNotFound:  ConnectErrorMsg = "File Not Found"
  Case moInvalidDirectory: ConnectErrorMsg = "Invalid Directory"
  Case moHostUnknown:  ConnectErrorMsg = "Unknown Host"
  Case Else:    ConnectErrorMsg = "Unrecognized Error Code"
 End Select
End Function
```

This code is used to handle data connection errors.

10 Run the program.

You should be able to add new shape files to the map.

Synchronizing the Map and the Add Layer Form

CD-ROM NOTE: *The VB project in the* Chapter5_2 *directory on the companion CD-ROM starts here.*

The current program adds the layer to the map control, but it does not update the list in form *Form2*. The strategy for having the program do so is as follows. After loading the new layer, delete the list of layers in form *Form2* and then reload the form. All of this should take place at the end of the *addFile* sub. To delete a list, you only need to call its *Clear* method. The following are the two lines you would add to the end of the *addFile* sub.

```
 lstLayers.Clear
 Form_Load
End Sub
```

Summary

In this chapter you worked with a VB control, the Common Dialog Control, and MO's *DataConnection* and *GeoDataset* objects. Using the Common Dialog Control's *ShowOpen* method, you added the capability for the user to load shape files found on a computer's disk to your mapping program. To implement this function, you had to parse the string returned by the File Open dialog into *DataConnection* and *GeoDataset* objects. Once the new shape was added to your program, you used the *Clear* method to delete the list of layers in form *Form2*. A final call to that form's *Form_Load* function ensured that the map layers and the list of layers were synchronized.

Chapter 6

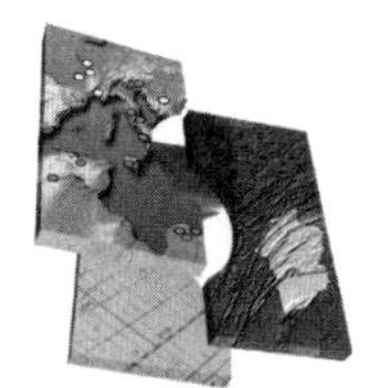

Toolbars and Layer Management

Introduction

If you think back to the last chapter, you started by looking at a VB issue (i.e., how to open a file using the *CommonDialog* object), and then studied how to connect that new tool to objects using MapObjects (MO). Specifically, you used the *FileOpen* dialog object to get a path. You also parsed the path to create a *DataConnection* object (from the path) and a *GeoDataset* object (from the file name). This chapter continues in this manner. That is, you will learn about new VB features and then look at how to connect them to MapObjects controls.

Adding a Remove Button

CD-ROM NOTE: *The VB project in the* Chapter6_1 *directory on the companion CD-ROM starts here.*

The first thing you will do is add a new button to the form containing the layer list (form *Form2*).

1. Add a button to the form.
2. Name this button *cmdRemove*, and give it a caption of *Remove*.

Can you guess what this does? It removes a layer from the map. Your form should look like that shown in figure 6-1.

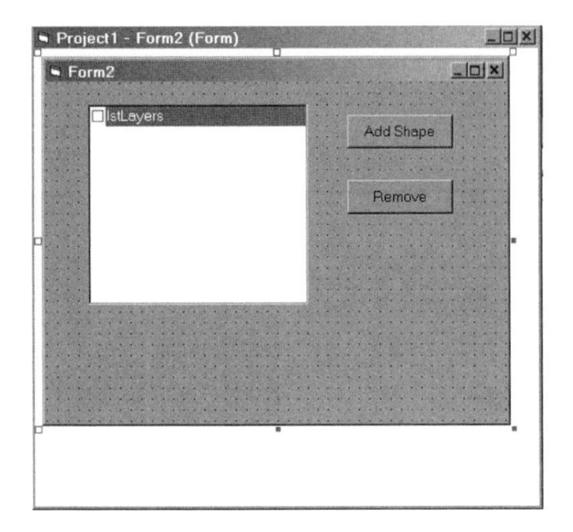

Fig. 6-1. Remove button added to form Form2.

What is required for this button to work? The user must select a layer, and then click on the Remove button. Should this button always be active? What if there are no layers in the map? In this case, you would want the button to be disabled. If the button is disabled and the user adds a layer, the button should be enabled. In other words, there are a lot of details to work out.

Let's begin with the actual process of removing layers. The button should do nothing if no layer is selected. How do you know if a layer is selected? What would be the value of *ListIndex*? (If you do not recall the meaning of *ListIndex*, refer to the discussion of list and combo boxes in Chapter 1.) The following is the structure of the code needed to remove a layer.

3 In form *Form2*'s code page, add the following.

```
Private Sub cmdRemove_Click()
 If        then
   Exit Sub
 End If
 'What do we know if we get to here?
 Form1.Map1.Layers.Remove(lstLayers.ListIndex)
 lstLayers.Clear
 Form_Load
 RefreshButtons
End Sub
```

You need to figure out what goes between the *If* and the *Then* to ensure that there is a layer to remove. What should happen when there are no layers in the list? In this case, the Remove button should be turned off. For this you will create a sub named *RefreshButtons*. Why do you need a sub? You could get by without a formal *RefreshButtons* sub. However, you need to continue to add buttons to this form. Therefore, a sub that contains all of the button-refreshing logic will prove useful. The following is the *RefreshButtons* sub. As you add more functionality to your program, you will add more cases to this sub.

4 Add the *RefreshButtons* sub to form *Form2*'s code page.

```
Private Sub refreshButtons()
  Dim curItem As Integer
  curItem = lstLayers.ListIndex
  'No items selected.
  If curItem = -1 Then
    cmdRemove.Enabled = False
  Else
    cmdRemove.Enabled = True
  End If
End Sub
```

This sub checks to see if there is any selected item. If a layer is selected, it can be removed, and the Remove button is enabled. Otherwise, the button is disabled. This tells you what should go between the *If* and *then* statements in *CmdRemove_Click* alluded to previously. You need to include a statement that checks the *ListIndex* property of *lstLayers*. That is, the first *If/Then* statement in the sub *CmdRemove_Click* should read as follows.

```
If lstLayers.ListIndex < 0 then
```

Where should calls to the *RefreshButtons* sub be made? Obviously, it must be called whenever a map layer is removed. What if a map layer is added? If the list were empty and suddenly got a layer, the Remove button would need to be enabled. This *RefreshButtons* sub accomplishes that.

There is another thing you must guard against when you remove layers. If there are no layers in the list, and the user clicks in the list box, this could fire an *ItemCheck* event. You do not want the *ItemCheck* sub to run when there are no layers in the list. Therefore, you need to place the following at the beginning of the *lstLayers_ItemCheck* sub.

5 Edit the *if* statement in *cmdRemove_Click* to read as follows.

```
if lstLayers.Count = 0 Then
  Exit Sub
End If
```

This example points out the difficulty in designing event-driven programs. If you did not place the foregoing code in your *itemcheck* function, your program would work just fine—until someone clicked on (checked) an empty list box!

NOTE: *The responsibility for thinking of all possibilities (such as the example of clicking on an empty list box) is the programmer's.*

Image Lists and Toolbars

Most computer users have used toolbars in a windowed environment. They are quite common. In fact, they are part of a component called the Microsoft Windows Common Controls. In this section, you will add these controls to this project.

1 Press Ctrl-T on your keyboard (see figure 6-2) to open the Add Components dialog. Add the Microsoft Windows Common Controls (ActiveX) component.

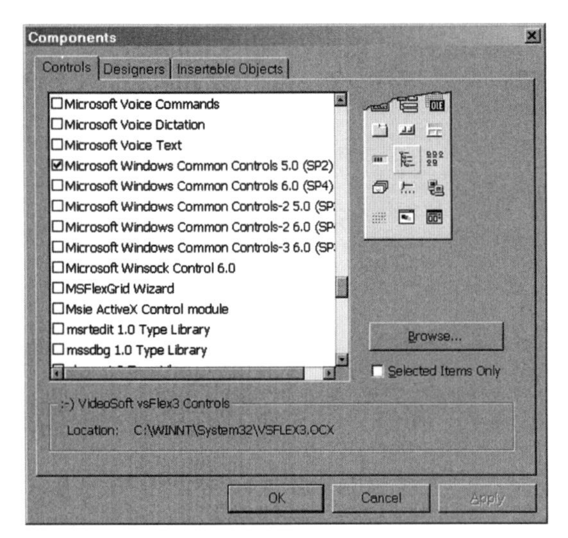

Fig. 6-2. Adding the Windows Common Controls component.

This ActiveX component provides you with some import widgets to add to a program. Let's first look at the Image List control. The Image List control is much like the Common Dialog control in that you place it on a form, but it is not visible at run time. With the Image List control, you place all images you need for a form in the list. This allows other components (in particular, the toolbar component) to use these images.

2 Place an Image List control (its icon looks like a stack of envelopes) on the form.

3 Right click on the Image List control to bring up its property form.

4 Click on the Images tab, and then on Insert Picture.

This will bring up an Open File dialog.

5 Navigate to the *Utility* directory on the companion CD-ROM. Add the images for Promote and Demote (see figure 6-3).

6 Place a toolbar on form *Form2.*

The ToolBar control was added to the toolbar when you incorporated the Microsoft Common Controls.

7 In the toolbar's properties box, make sure the Align value is set to *vbAlignNone.*

8 Name the toolbar *layerTools.*

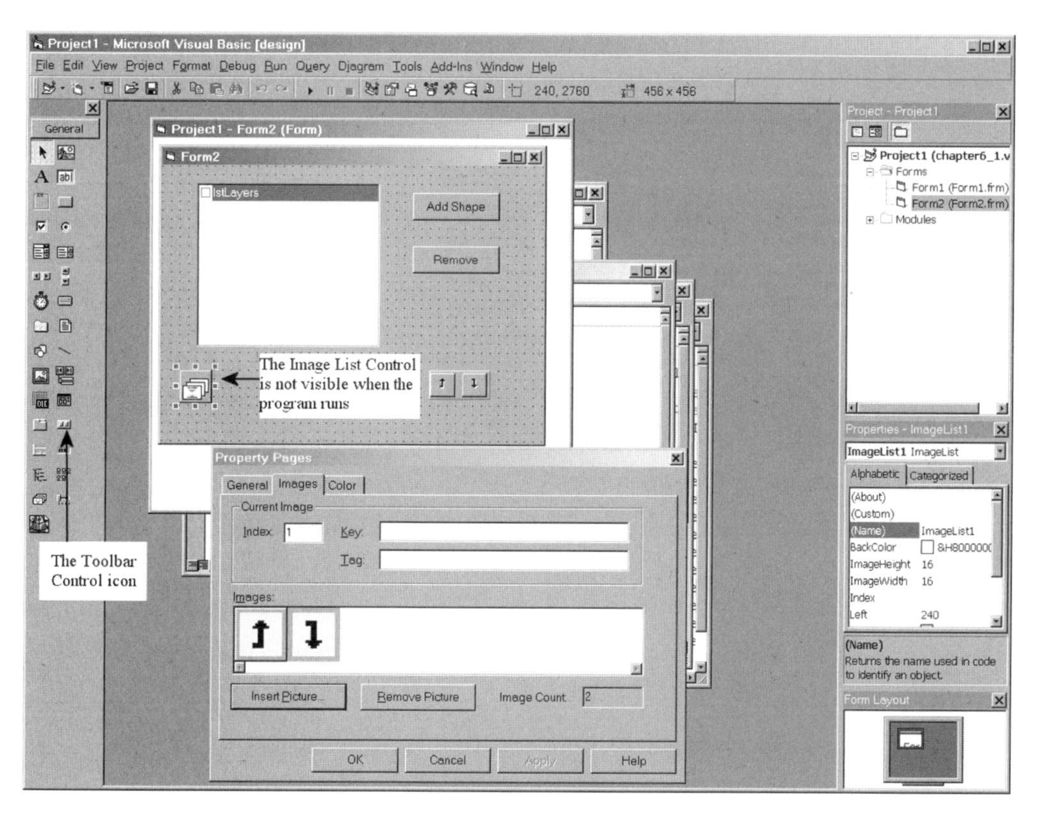

Fig. 6-3. Promote and Demote images added to the Image List control.

9 Drag the toolbar to near the bottom of form *Form2.*

The next phase involves adding buttons to the toolbar and assigning the images in the Image List control to those buttons.

10 Once you have the toolbar located and sized the way you want, right click on it. This will bring up another properties box.

11 In the property pages for the toolbar, perform the following:

- ❐ On the General Tab page, set the *ImageList* option to *ImageList1* (the list you created that contains the Promote and Demote images).
- ❐ Click on the Buttons tab. Insert a button. Set its description to *Promote,* its key to *Promote,* its value to *0-tbrUnpressed,* its style to *0-tbrDefault,* its tool tip text to *Move Layer Up,* and its image to *1* (the first picture you added to the list).

Your toolbar properties form should look like that shown in figure 6-4.

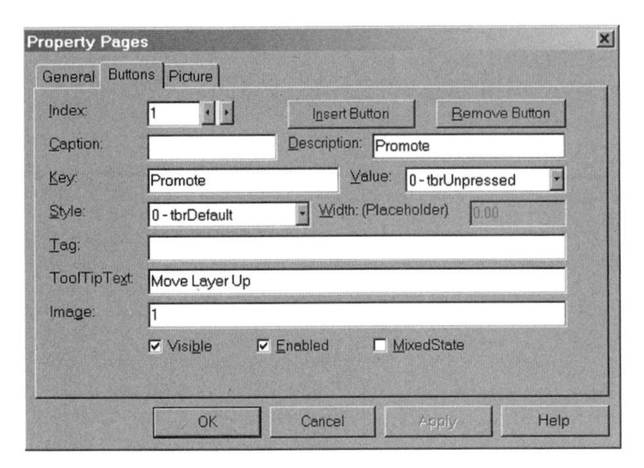

Fig. 6-4. Revised toolbar properties form.

12 Insert a second button. Set its Style to *3-tbrSeparator.*

13 Insert a third button. Set its description to *Demote,* its key to *Demote,* its value to *0-tbrUnpressed,* its style to *0-tbrDefault,* its tool tip text to *Move Layer Down,* and its image to *2* (the second picture you added to the list).

14 Dismiss the property pages.

You now have a toolbar on form *Form2.* You may have to resize it to get the images to display properly.

15 Double click on the toolbar.

This will place the private sub *layerTools_ButtonClick* in form *Form2*'s code page.

16 Add the following code.

```
Private Sub layerTools_ButtonClick(ByVal Button As ComctlLib.Button)
 Dim curIndex As Integer
 curIndex = lstLayers.ListIndex
 'Here, we promote or demote the layer.
 Select Case Button.Key
 Case "Promote"
   Form1.Map1.Layers.MoveTo curIndex, curIndex - 1
   lstLayers.Clear
   Form_Load
   lstLayers.ListIndex = curIndex - 1
 Case "Demote"
   Form1.Map1.Layers.MoveTo curIndex, curIndex + 1
   lstLayers.Clear
   Form_Load
   lstLayers.ListIndex = curIndex + 1
 End Select
 refreshButtons 'Refresh the button enabled status on Map Contents
 Form1.Map1.Refresh 'Redraw the map with the new Layer order
End Sub
```

What does this function do? It moves the layers up or down. The *MoveTo* function is an MO function. The calls to *lstLayers.Clear* and

From_Load rebuild the list to reflect the new drawing order. The last two calls refresh the buttons and redraw the map.

The *RefreshButtons* function was a simple concept when you only had to worry about one button (the Remove button). Now, however, you have two more buttons: the Promote and Demote buttons. You have to decide when these buttons should be turned off or on. You need to answer the following questions.

- ❐ If there are no layers in the map, what should be the status of these buttons?
- ❐ If there is only one layer in the map, what should be the status of these buttons?
- ❐ If no map layer is selected (*listIndex = -1*), what should be the status of these buttons?
- ❐ If the top map layer is selected, what should be the status of these buttons?
- ❐ If the bottom map layer is selected, what should be the status of the buttons?
- ❐ If there is more than one map layer, and the selected layer is neither the top nor the bottom layer, what should be the status of these buttons?

One last question. What if someone selects a layer (a *click* event) but does not create an *itemcheck* event; what should happen? This occurs when the user clicks on the layer name, but not in the check box.

You need to add a variable to *RefreshButtons* that will refresh buttons and keep track of the number of layers in the list. Let's name this variable *listCount.* The first (and easiest) case to consider is when no item is selected. The following code shows the implementation of the *listCount* variable.

17 Edit *RefreshButtons* as follows.

```
Private Sub refreshButtons()
 Dim listCount As Integer
 Dim curItem As Integer
 listCount = Form1.Map1.Layers.Count
 curItem = lstLayers.ListIndex
 'No items selected.
 If curItem = -1 Then
   cmdRemove.Enabled = False
```

```
    layerTools.Buttons("Promote").Enabled = False
    layerTools.Buttons("Demote").Enabled = False
```

The previous code manages the case when no item is selected (*curItem = –1*). However, if *curItem* is not equal to –1, an item is selected. The program should check to see if only one item is in the list. If this is the case, the program can neither promote nor demote the map layer. The following code manages the case when only one item is in the list (i.e., there is only one map layer).

18 Continue editing *RefreshButtons* as follows.

```
  'Only one item in list.
  ElseIf listCount = 1 Then
    cmdRemove.Enabled = True
    layerTools.Buttons("Promote").Enabled = False
    layerTools.Buttons("Demote").Enabled = False
```

If *listCount* is greater than 1, a layer is selected and more than one layer is in the list. The program then needs to check if it is the first (top) layer, last (bottom) layer, or a layer somewhere between these two. The following code accomplishes this.

19 Finish editing *RefreshButtons* as follows.

```
  'Many items, see if first item is selected.
  ElseIf curItem = 0 Then
    cmdRemove.Enabled = True
    layerTools.Buttons("Promote").Enabled = False
    layerTools.Buttons("Demote").Enabled = True
  'Many items, see if last item is selected.
  ElseIf curItem = listCount - 1 Then
    cmdRemove.Enabled = True
    layerTools.Buttons("Promote").Enabled = True
    layerTools.Buttons("Demote").Enabled = False
  'Many items, an item between first and last
  'is selected.
  Else
    cmdRemove.Enabled = True
    layerTools.Buttons("Promote").Enabled = True
    layerTools.Buttons("Demote").Enabled = True
  End If
End Sub
```

The *RefreshButtons* sub now contains code for managing the state of the Promote and Demote buttons on form *Form2*'s toolbar. You need to do one last thing.

20 Add a call to *RefreshButtons* to the end of *Form_Load*. This will ensure that the Promote and Demote buttons are in the correct state when the form first loads.

21 Run this program. You should be able to add and remove layers, and to change the drawing order.

The program should work fine.

The Map Toolbar

CD-ROM NOTE: *The VB project in the* Chapter6_2 *directory on the companion CD-ROM starts here.*

If you can place a toolbar on form *Form2*, you can, obviously, place one on form *Form1*. In this section you will develop a toolbar for managing the user's interaction with the map.

1 Add an Image List control and a toolbar control to form *Form1*.

You then need to populate the Image List with the following images, found in *C:\Program Files\ESRI\MapObjects\Samples\VB\MoView2\Bitmaps.*

- *mapcontents*
- *zoomin*
- *zoomout*
- *pan*
- *identify*
- *fullextent*

2 Add six buttons to the toolbar. Assign each button an image from the images added in step 1. Do this in the order in which the images were previously listed (see figure 6-5).

3 Give the buttons added in step 2 the following key values and styles. Set the description equal to the key value. (The key value is listed first, followed by the style.)

- LayerControl, tbrDefault
- ZoomIn, tbrButtonGroup

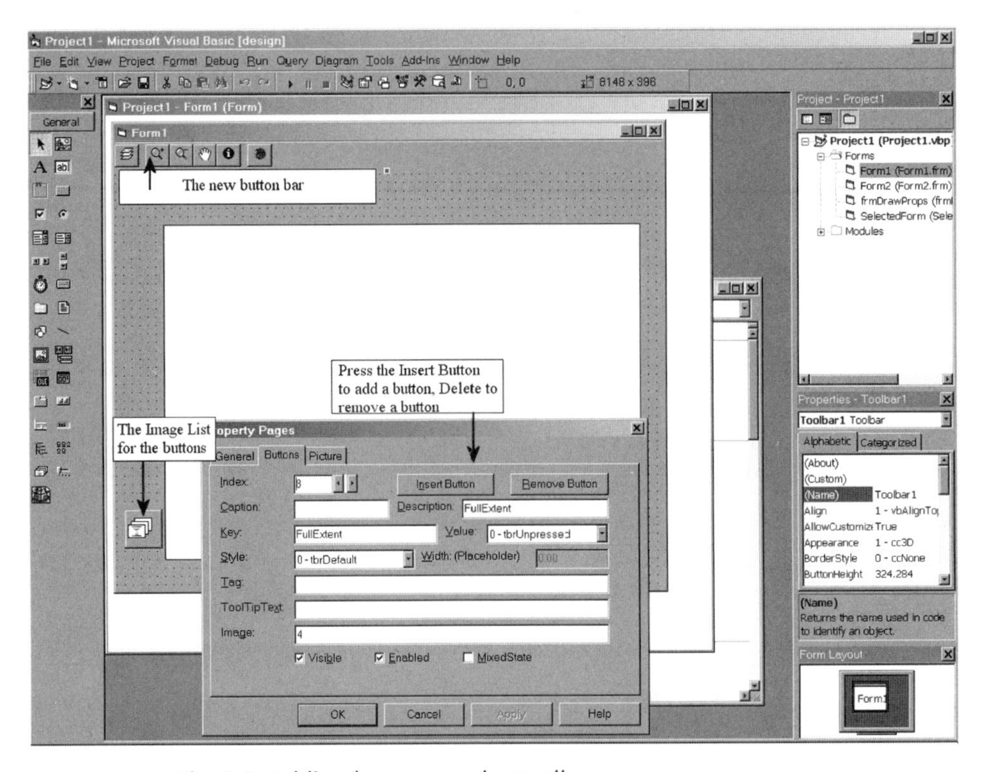

Fig. 6-5. Adding buttons to the toolbar.

- ❐ ZoomOut, tbrButtonGroup
- ❐ Pan, tbrButtonGroup
- ❐ Identify, tbrButtonGroup
- ❐ FullExtent, tbrDefault

4. Double click on the toolbar. This adds a *ButtonClick* event sub to form *Form1*'s code page.
5. Add the following code to this subroutine.

```
Select Case Button.Key
Case "LayerControl"
  Form2.Show
End Select
```

6. Remove the *cmdLayers* command button and its *click* event from your project.

In the remainder of this chapter and in subsequent chapters, you will add the functionality to each button.

Zoom Methods

To implement the zoom methods (functionality previously discussed), you need to consider a new object: the *rectangle* object. The *rectangle* object has several properties and methods. You will work with the following.

- *Properties:* Top, Left, Bottom, Right
- *Methods:* Intersect, Union, ScaleRectangle

The map control has two properties you need to consider: *Extent* and *FullExtent.* The former is the current extent of the map. The latter is the extent that would contain all layers.

Implementing the Full Extent button (the last one on the button bar) is straightforward.

1. To the *ButtonClick* sub, add the following case before the *End Select* statement.

```
Case "FullExtent"
  Map1.Extent = Map1.FullExtent
```

The Zoom In button works a little differently. When someone presses the Zoom In button, nothing happens until they indicate the zoom area by drawing a rectangle on the map. The Map control has a *TrackRectangle* method. You track the rectangle (get its extent) and set the map extent to that of the rectangle. Tracking the rectangle occurs when there is a *MouseDown* event in the map control.

2. Go to form *Form1*'s code window. In the left-hand drop box, select *Map1.* In the right-hand drop box, select *MouseDown.*

This will add an empty sub, as follows.

```
Private Sub Map1_MouseDown(Button As Integer, Shift As Integer, _
      X As Single, Y As Single)
End Sub
```

3. Add the following code to the *MouseDown* sub.

```
If Toolbar1.Buttons("ZoomIn").Value = 1 Then
  Map1.Extent = Map1.TrackRectangle
  Map1.Refresh
End If
```

The *If* statement asks if the Zoom In button (*"ZoomIn"*) is pressed. If so, when the mouse is down, the rectangle is traced. Once the mouse is up, the map extent is set to the rectangle extent and the map is refreshed.

Users of ArcView 3 are familiar with the difference between buttons and tools. Buttons execute immediately (such as the Full Extent button and the Layers button). Tools require further input from the user (such as for the Zoom In button). Here you have mixed the two types of controls in one toolbar. If you wanted to emulate ArcView 3, you could have created two toolbars: one as a set of tools and the other as a series of buttons. In adjunct exercise 6-1, which follows, you have the opportunity to practice working with pan and zoom functionality.

Adjunct Exercise 6-1: Working with Pan and Zoom

The objective here is to get the Zoom Out and Pan buttons to work. Both of these buttons will be in *Map1_ MouseDown* events. The Pan button uses the Map control's *Pan* method (very easy to implement). The Zoom Out button is a bit more complex. Look at MoView for some ideas. If you have trouble doing this, look at the code in the directory *Chapter6_2* on the companion CD-ROM.

Summary

Toolbars are part of Microsoft Windows Common Controls. In this chapter you added a toolbar to both forms in your project. You also used the Image List control to add images that could be used as icons on buttons you placed on the toolbars. The toolbar in form *Form2* managed the drawing order of layers. You associated clicks on the Promote and Demote buttons with the map layers collection object's *MoveTo* method. Proper management of this toolbar required adding code to the *RefreshButtons* sub so that the buttons on the toolbar were only enabled when layers could be promoted or demoted.

The toolbar you added to form *Form1* contained buttons that provoke an immediate action, and some that require further input from the user. Those that require an immediate action are managed by the toolbar's *ButtonClick* event subroutine. Those that require more input from the user are managed by the map control's *MouseDown* event subroutine.

Chapter 7

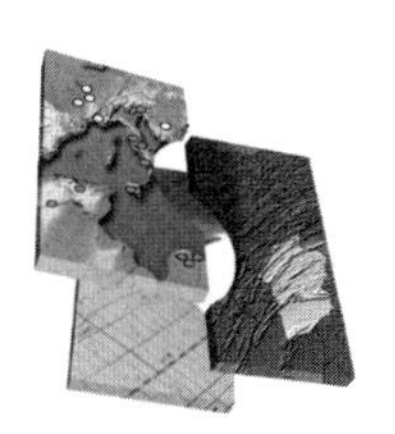

Geometry, Coordinates, and Identifying Features

Introduction

In Chapter 6 you added a toolbar to form *Form1* and several buttons. One of those buttons was the Identify button. In this chapter you will implement that button. For the Identify button to work, you need to be able to convert mouse clicks on the map to map units, and then determine what features (if any) are near the mouse click. To accomplish these tasks, it is necessary to use the distance methods in MapObjects (MO).

Distance Methods

MapObjects projects can use shapes from various areas. We might use shapes of the world (countries, lakes, and so on), shapes of a particular country, or shapes of local areas. The sample shapes that come with MO and the shapes in the *USA* directory on the companion CD-ROM are in latitude and longitude. However, computer screens do *not* display in map units (e.g., latitude/longitude). Screen units are referred to as "twips." Therefore, MO incorporates functions that convert screen units (twips) to map units (latitude/longitude, meters, feet, or whatever units the shapes are in) and map units to screen units. Look at the map-display object diagram (figure 7-1).

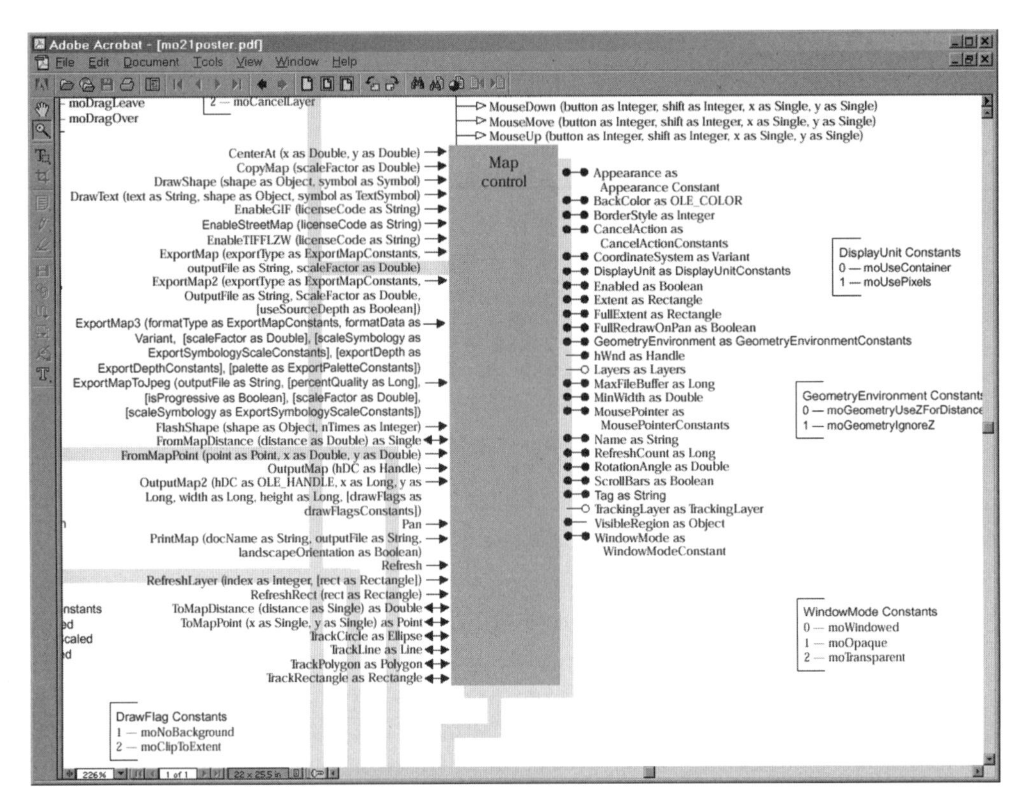

Fig. 7-1. Map display object diagram.

The Map control has a number of functions, four of which perform coordinate transformations: *FromMapDistance*, *FromMapPoint*, *ToMapDistance*, and *ToMapPoint*. The following is an example you can try with a *MouseDown* event on the Map control.

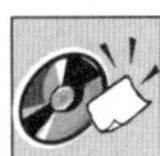

CD-ROM NOTE: *The VB project in the* Chapter7_1 *directory on the companion CD-ROM starts here.*

```
Dim ln As MapObjects2.Line
Set ln = Map1.TrackLine
MsgBox "Map Units: " & ln.Length & vbCr & _
"Control Units: " & Map1.FromMapDistance(ln.Length)
```

In this section of code, the MO function *TrackLine* is used to keep track of a line. The line is finished when you double click. At that time, the *MsgBox* function displays the map units in latitude/longitude and in twips. The twips are calculated by the function *FromMapDistance*. *ToMapDistance* does just the opposite. It converts from twips to map units.

The point functions *FromMapPoint* and *ToMapPoint* convert map units to twips and twips to map units, respectively, for a given point location. The following is an example of how you might use them in a zoom-out function.

Suppose the user clicks on the map. The program needs to obtain the *x-y* coordinates of that location (where the user clicked) in twips. Note that the *MouseDown* event reports the current *x* and *y* locations. You then create a rectangle and set its extent to the current map extent. Next, you scale the rectangle using the geometric function *ScaleRectangle* (see the geometric object's diagram) by two.

Finally, you set the map extent to the newly scaled rectangle and then center the map on the coordinates returned by the *ToMapPoint* function. Adjunct exercise 7-1, which follows, provides you with opportunity to practice implementing a zoom-out function using the *ScaleRectangle* and *ToMapPoint* methods.

Adjunct Exercise 7-1: Implementing the Zoom-out Function

Implement the zoom-out function as described previously (not as in *MoView2*).

Resizing the Main Form

When you work in twips, 0,0 is at the upper left. Y values increase downward, not upward, in the control. At the moment, your main form has two controls on it: a toolbar control and a map control. The height of the toolbar control is returned by the *Toolbar1.Height* request. If you had menus, the top of the control would be returned by the *Toolbar1.Top* request. These requests (*Get* functions) are VB requests; that is, they are not unique to MO. Two other VB requests you will use are *ScaleWidth* and *ScaleHeight*. These measure the interior size of an object in twips. Consider what would happen were you to resize a form.

Obviously, the form changes size. That is, the *ScaleWidth* and *ScaleHeight* of the form take on new values. VB looks for a *Form_Resize* sub to handle such an event and to determine the size and location of the controls on the form. Your form currently contains the toolbar and map controls. The toolbar cannot resize with the form, but the map control can. In the *x* direction, the map control should run from the left side of the form to the right side of the form, or from 0 to *ScaleWidth.*

The *y* dimension is a bit trickier. The map control should run from the bottom of the toolbar to the bottom of the form. The entire extent of the form in the *y* direction is *ScaleHeight,* but the map control must leave room for the toolbar on the form. Therefore, the height of the map control should be the *ScaleHeight* of the form minus the toolbar height. The following *Form_Resize* sub (added to form *Form1*) achieves this.

```
Private Sub Form_Resize()
 Dim mapTop As Integer
 mapTop = Toolbar1.Top + Toolbar1.Height
'note that 0,0 is the upper right
 Map1.Move 0, mapTop, ScaleWidth, ScaleHeight - mapTop
End Sub
```

In regard to other geometric features, as you can see from the MapObjects object diagram, there are several geometric objects: the point object, the points collection object, the line object, the polygon object, the rectangle object, and the ellipse object. In addition, there is the parts collection. Many of these are discussed in depth later in the book.

The MapObjects Recordset Object

Identifying features on a map requires retrieving the records associated with the selected features. The *Identify* function should allow the user to select features with the mouse, and to retrieve the attributes of those features and report them to the user. These abilities require the use of the MO *Recordset* object. You must understand the *Recordset* object before you can implement the *Identify* function in your VB project.

You have to be careful when declaring a *Recordset* object. Both MO and VB have *Recordset* objects (recall the bookstore example of Chapter 2). To declare an MO *Recordset* object, use the following format.

```
Dim recs as New MapObjects2.Recordset
```

Think of a *Recordset* object as a table. The table has a set of fields, and a current row. Look at the object diagram for record sets (figure 7-2).

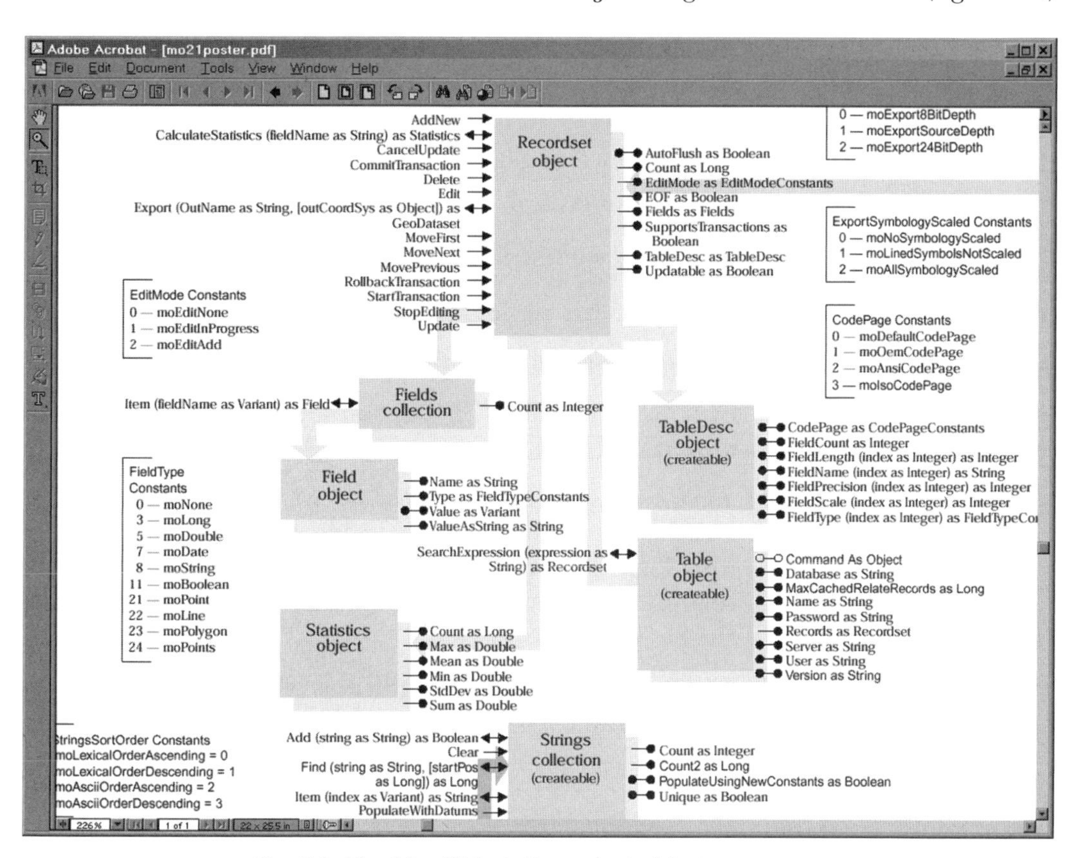

Fig. 7-2. The MapObjects Recordset *object.*

Every map layer has a *Recordset* object associated with it. If you select certain features from a map layer (discussed later in the book), the selected set of that layer's features becomes a *Recordset* object as well. Dealing with features in a map layer consequently requires that you deal with MO record sets. Therefore, an understanding of some useful methods, properties, and events associated with MO record sets is important. The following are a few of the methods you will find useful.

- *AddNew:* Adds a new record to the bottom of a record set
- *CalculateStatistics:* Calculates statistics for a specified field in the record set
- *Delete:* Deletes the current record from the record set

- *Edit:* Allows edits of the current record
- *MoveFirst:* Makes the first record in the record set the current record
- *MoveNext:* Makes the next record in the record set the current record
- *MovePrevious:* Makes the previous record in the record set the current record

The following are some commonly used properties of record sets.

- *Fields:* Returns all fields associated with the record set.
- *Count:* Returns the number of records in the record set.
- *EOF:* Returns TRUE if the pointer to the record set has moved past the last record.
- *TableDesc:* Returns the *TableDesc* for the record set. The *TableDesc* is the list of fields found in a standard database file. That is, it does not contain information about the relationship between the map and the database, such as the Shape field.
- *Updatable:* This property indicates whether the record set can be edited.

Adjunct Exercise 7-2: Working with Record Sets

To practice working with record sets, perform the following.

1. Load a layer into your map.
2. Get its record set, and its number of fields (*aRecordset.Fields.Count*).
3. Declare *tdesc* a *New MapObjects2.TableDesc.*
4. Set *tdesc = aRecordset.TableDesc.*
5. Find the layer's number of fields (*tdesc.FieldCount*).

You should note that *aRecordset* has two more fields than *tdesc.* The reason for this is that the *Recordset* object contains the fields that indicate the relationship between the layer's database table and the layer's geometry. For shape files, these two extra fields are named *FeatureId* and *Shape.*

Managing the Identify Button

You can use some of the *Recordset* object's properties and methods in creating the Identify button. If the user selects a set of features, a new record set (which we will call *gSelection*) is created. To display the record set, the program will present a form that lists the selected features and their table values. This might seem like an easy task, but the devil is in the details. Let's begin with the easy tasks.

1. On the main form, resize the map control to leave some space between it and the toolbar.
2. Add a combo box below the toolbar. Give it a label with the caption *Current Layer.* Give the combo box a style value of *2-drop down.*

Your form should now look like that shown in figure 7-3.

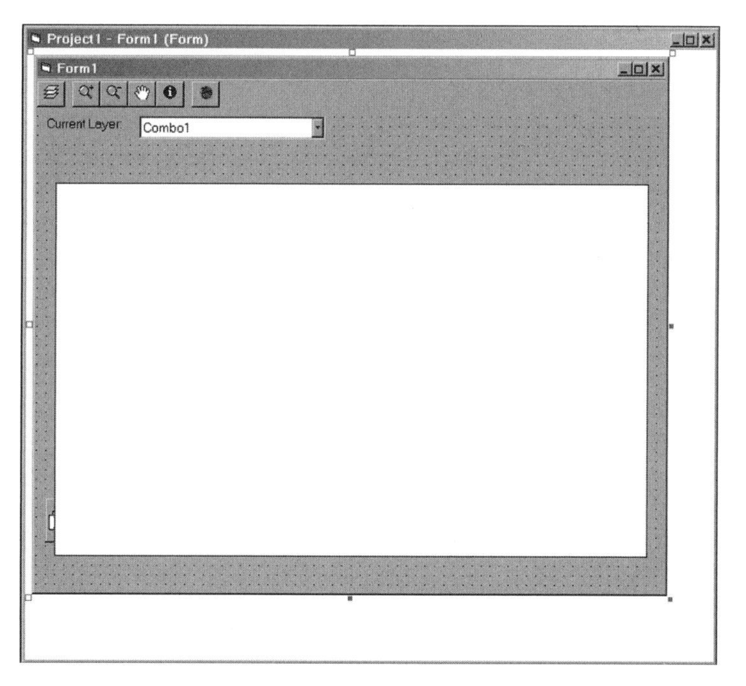

Fig. 7-3. Form Form1 *with the Current Layer combo box.*

The combo box will allow the user to choose which layer is the current layer; that is, which layer will be queried when the Identify button is selected. Clearly, you have to synchronize the content of the combo box with the layers in the map. If there are no layers in the map, there should be no content in the combo box. If the layer order is changed, it probably makes sense to change the layer order in the combo box. Before jumping into this, however, you need to make sure the map control does not cover up the combo box.

3 Change the *mapTop* line in the *Form_Resize* function to:

```
mapTop = Label1.Top + Label1.Height + 75
```

The 75 twips should provide a sufficient visual separation between the map control and the combo box. The selected item in the combo box will store the name of the current (active) layer. You need a variable for storing this layer. This variable will need to be visible to at least two forms: the main form and the form that displays information about the selected features. Therefore, it should be global in scope.

4 Input the following in the *modUtility* module.

```
Public ActiveLayer As MapObjects2.MapLayer
```

When the user clicks on the combo box, *ActiveLayer* needs to be set equal to the layer the user has chosen.

5 To the combo box's click event, add the following:

```
Private Sub Combo1_Click()
 If Combo1.ListIndex >= 0 Then
  Set ActiveLayer = Map1.Layers(Combo1.ListIndex)
  Toolbar1.Buttons("Identify").Enabled = True
  'This works but we must keep the combo list
  'and the map layers synchronized
 End If
End Sub
```

Why is the *If* statement necessary here? Suppose there are no layers and the user clicked on the empty combo box. What would happen? The variable *Combo1.ListIndex* would be –1. If the program tried to find a map layer with an index value of –1, the program would crash.

Note that the Identify button is enabled in this sub as well. Where should it be disabled? That is, when can there be no active layer? The following are some possibilities. Suppose you started the program with an empty map (i.e., with no map layers). Clearly, the Identify button should be disabled. Suppose the button were enabled but the active layer removed (recall the *Remove* function in form *Form2*) from the map. In this case, there would be layers in the map, but none of them would be the active layer.

Let's consider other things that could affect the combo box. If you add layers to the map, the list for the combo box should change. The same is true if you delete layers from the map. However, if you have an active layer and add a new layer to the map, the status of the Identify button should not change. If you delete a layer that is not the active layer, the list changes, but the selected item and the Identify button status should remain the same. It looks like you will have to update the combo box whenever there is a change to the map control's layers. The following code achieves this.

6 Add the following code to form *Form1*'s code page.

```
Public Sub RefreshCombo1()
 Dim curselected As String
 If (Combo1.ListIndex >= 0) Then
  curselected = Combo1.List(Combo1.ListIndex)
  Toolbar1.Buttons("Identify").Enabled = True
 Else
  curselected = ""
  Toolbar1.Buttons("Identify").Enabled = False
 End If
 Combo1.Clear 'Destroy the list's current contents
 If Map1.Layers.Count = 0 Then
  Exit Sub
 End If
 For i = 0 To Map1.Layers.Count - 1
  Combo1.AddItem Map1.Layers(i).Name
  If Combo1.List(i) = curselected Then
      Combo1.ListIndex = i
      Set ActiveLayer = Map1.Layers(i)
      End If
 Next i
End Sub
```

This is a general function (not tied to any form) because no control directly triggers it. It is also public, and therefore you can call it from other forms when necessary. The *RefreshCombo1* sub works as follows. It starts by saving the current *ActiveLayer* name, if there is one, and setting the Identify button accordingly. It then clears the content of the combo box. If there are no layers in the map, the sub ends. If there are layers in the map, the last *For* loop adds each layer's name to the combo box, sets the value of *ActiveLayer* equal to the currently selected layer, and sets the *ListIndex* for the combo box to the selected layer.

When should you call this sub? Let's assume that no layer is the active layer when the program starts. Therefore, the sub should be called from form *Form1*'s *Load* sub. Clearly, when you delete or add layers, or when you change their order, you need to call this sub. The following are the locations from which you need to call this sub.

7. In form *Form1*'s code, this is easy. Simply put a call to *RefreshCombo1* in the *Form_Load* function.
8. Add the following to the bottom of form *Form2*'s *RefreshButtons* sub.

```
Form1.RefreshCombo1
```

You need to place the call in form *Form2*'s *RefreshButtons* sub. Let's see why. Currently, form *Form2*'s *RefreshButtons* sub is called when the form is loaded, when layers are promoted or demoted, when layers are added or removed, and when the user clicks on the list of layers. With the exception of the first and last cases, you also call the *Form_Load* function. That is, when you promote, demote, add, or remove a layer, the *RefreshButtons* sub is called *twice*.

Now consider what happens when you remove a layer that is the active layer. The first call to *RefreshButtons* (from form *Form2*'s *load* function) makes a call to *RefreshCombo1*. The *ListIndex* value is greater than or equal to zero. Therefore, the following *If* statement in *RefreshCombo1* is executed.

```
If (Combo1.ListIndex >= 0) Then
  curselected = Combo1.List(Combo1.ListIndex)
  Toolbar1.Buttons("Identify").Enabled = True
```

The sub continues and clears the combo box list. This automatically sets *ListIndex* to –1. The sub moves on to the *For* loop (assuming there are still layers in the map control). As it loops, it cannot find a layer name to match that stored in *curselected*. Therefore, *ListIndex* remains –1. Control is passed back to form *Form2*. At this point, the combo box has no selected item, but the Identify button is still active.

The second call to *RefreshButtons* is made, and this fires a second call to *RefreshCombo1*. This time, *ListIndex* is –1. Therefore, the following code is executed.

```
Else
  curselected = ""
  Toolbar1.Buttons("Identify").Enabled = False
```

This turns off the Identify button, which is what you want. The point is that if you had processed form *Form2* in some other way (without two calls to *RefreshButtons*), the *RefreshCombo1* function would not have worked.

At this point, you have the loading and maintenance of the combo box working, and the status of the Identify button working. You now need to specify what happens when the user clicks on the Identify button.

Retrieving a Selection

Now that you have the Identify button working and synchronized with the active layer, you need to add the elements to make the button work. Clicking on the Identify button is like clicking on a zoom-in, zoom-out, or pan button: it takes a second action (a *MouseDown* action) to complete the *Identify* function. When the user selects the features she wants to identify, the program needs to highlight those (draw them in a specific color) and then display a form containing the appropriate information. Let's first focus on selecting features. You will store selected records in a global *Recordset* object named *gSelection*.

1. In *modUtility*, create a public *MapObjects2.Recordset* and name it *gSelection*. That is, the start of *modUtility* should appear as follows.

   ```
   Option Explicit
   Public gSelection As MapObjects2.Recordset
   Public ActiveLayer As MapObjects2.MapLayer
   ```

You now need to insert code in form *Form1*'s *MouseDown* event to handle the Identify button. There are two methods by which a user could select features to identify. The first would be to draw a rectangle and select all features that intersect with that rectangle. The second method would be to simply point at a feature and click the mouse.

2. Return to the *MouseDown* event in form *Form1*'s code and add a section for the Identify button. After the last line in the Pan tool's code (*Map1.Refresh*), add the following code.

```
ElseIf Toolbar1.Buttons("Identify").Value = 1 Then
  Dim r As Rectangle
  Dim recs As MapObjects2.Recordset
```

```
  Dim tol As Double
  Set r = Map1.TrackRectangle
  Set recs = ActiveLayer.SearchShape(r, moAreaIntersect, "")
  If recs.EOF Then
   tol = Map1.ToMapDistance(100)
   Set pt = Map1.ToMapPoint(X, Y)
   Set recs = ActiveLayer.SearchByDistance(pt, tol, "")
  End If
If recs.Count > 0 Then
  Set gSelection = recs
Else
  Set gSelection = ActiveLayer.SearchExpression("featureId = -1")
End If
Map1.Refresh
```

The first section of this code creates and tracks a rectangle. Then the spatial search function *SearchShape* is used to see if any features of *ActiveLayer* intersect the rectangle. If *recs.Count* is 0 (the rectangle did not intersect any features), the program converts the mouse click to map coordinates and then checks to see if any features in *ActiveLayer* are within a specific distance of the mouse click. In the previous code, the search distance (*tol*) is set to 100 twips.

If *recs* is greater than 0 (the user selected something), *gSelection* needs to be set to that record set. If *recs.Count* equals 0 (nothing was selected), *gSelection* needs to be set to an empty set. That is what the last *If/Then-Else* statement in the previous code does.

The *Identify* section of the *MouseDown* event ends with a call to *Map1.Refresh.* The map is redrawn because you want to give the user some visual representation of the selected set. That is, you want the selected shapes drawn in a special color. To do this, you will use the *DrawShape* function in MO. This function can be called only by specific events. The specific event you will use here is *AfterLayerDraw.*

3 In form *Form1*'s code page, create the *AfterLayerDraw* sub for the *Map1* object. The easiest way to do this is to set the Object combo box at the top of the code page to *Map1* and the corresponding procedure to *AfterLayerDraw.* This will add the following sub to the form.

```
Private Sub object_AfterLayerDraw(ByVal index As Integer, ByVal _
   canceled As Boolean, ByVal hDC As Stdole.OLE_HANDLE)
End Sub
```

4 Add the following code so that the *AfterLayerDraw* sub is as follows.

```
Dim sym As New MapObjects2.Symbol
If index > 0 Then
      Exit Sub
End If
If gSelection Is Nothing Then
      Exit Sub
End If
 sym.Color = moRed
 gSelection.MoveFirst
 Do While Not gSelection.EOF
  Map1.DrawShape gSelection("Shape").Value, sym
  gSelection.MoveNext
 Loop
```

Let's consider the various components of the *AfterLayerDraw* sub. This sub is called after each layer is drawn. The *index* value represents the index of the layer most recently drawn. In this case, you want the selected shapes redrawn *after* the topmost layer is drawn. Recall that the index value of the topmost layer is 0. Thus, the *AfterLayerDraw* sub begins by checking if *index = 0*. If it does not, the program exits the sub.

If all layers have been drawn, *index = 0*. The sub then checks to see if there are any selected features to be drawn. It does this by checking if *gSelectoin* is nothing. If this is the case, the program exits the sub. If the program gets past the first two *If/Then-End If* statements, all layers have been drawn and there are selected records to be identified. The selected records will be drawn in red. This is set by the statement *sym.Color = moRed.*

The program then moves to the first element in *gSelection* and using a *Do-While* loop draws each selected feature with the proper symbol color. The drawing is accomplished by using the *DrawShape* function on the current record in *gSelection* using the symbol *sym*. Notice how the geometry of the shape is passed to the *DrawShape* function. The statement *gSelection("Shape").Value* returns the geographic coordinates and shape type (point, line, or polygon) to the *DrawShape* function.

Displaying Selected Records

CD-ROM NOTE: *The VB project in the* Chapter7_2 *directory on the companion CD-ROM starts here.*

At this point the code you have created will load and manage the combo box, select an active layer, set the Identify button's status, allow the user to select a set of features, and draw features with a special selection color. The code now needs to display the data for the features the user selects. Continue with the following.

1. Add a new form to your project and name it *Selected Form.*
2. Add the following controls to the new form.

- A text label called *Numfound*
- A text label called *ftypeLabel*
- A text label called *featurePrompt,* with the caption *Current Feature ID*
- A combo box called *cboIDs*
- A list box called *fieldList*

The new form should look like that shown in figure 7-4.

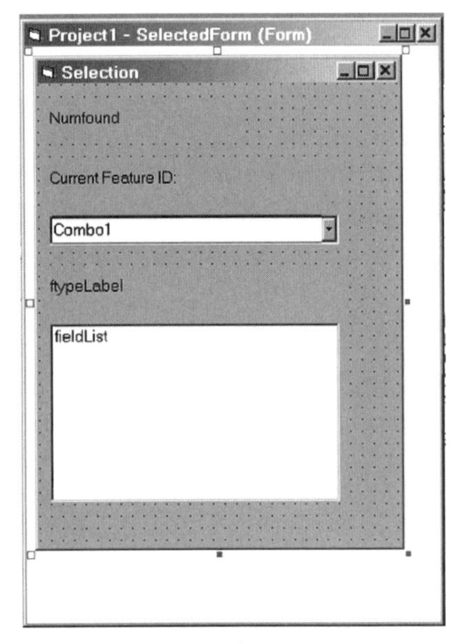

Fig. 7-4. Selection form.

Each control will be used as follows. *Numfound* will list the number of items in *gSelection.* The type of feature (point, line, or polygon) will be reported in *ftypeLabel.* The combo box will contain a list of the feature IDs of all elements in *gSelection.* The field list will contain all fields and their values for the current feature. The current feature is the one whose ID is displayed in *cboIDs.*

Adding items to the combo box list causes VB to fire a click event. You need to have a variable that will distinguish between this "simulated" click event and a "real" click event (one that occurs when the user clicks on the list). Recall the *FormUp* Boolean of form *Form2.* That variable was used to distinguish between instances of layer names being added to a list and actual mouse clicks on the list. You will use a similar strategy here, involving a Boolean named *SelectedFormUp.*

However, you will need to make this Boolean global, because it must be seen by *Form1* and *SelectedForm*.

3 Add the following to *modUtility*.

```
Public SelectedFormUp As Boolean
```

In the *SelectedForm*'s code, you will need an *Unload* function and a *cboIDs_Click* function. The *Unload* function should clear the selected set and refresh the map.

4 Add the following to *SelectedForm*'s code page.

```
Private Sub Form_Unload(Cancel As Integer)
Set gSelection=ActiveLayer.SearchExpression("featureId=-1")
 Form1.Map1.Refresh
End Sub
```

5 Add the *cboIDs Click* function to *selectedForm*'s code page, as follows.

```
Private Sub cboIDs_Click()
 Dim curfield as MapObjects2.Field
 Dim I as Integer
 If SelectedFormUp Then
   Exit Sub
 End If
 gSelection.MoveFirst
 fieldList.Clear
 For i = 1 To cboIDs.ListIndex
   gSelection.MoveNext
 Next
 For Each curfield In gSelection.Fields
   If curfield.Type = moString Then
      fieldList.AddItem curfield.Name + " = " + curfield.Value
  Else
      fieldList.AddItem curfield.Name + " = " + curfield.ValueAsString
  End If
 Next curfield
 Form1.Map1.FlashShape gSelection("shape").Value, 3
End Sub
```

This sub will behave as follows. If the Boolean *SelectedFormUp* is true, the program simply leaves the sub. If this Boolean is *not* true, the user clicked on the list of IDs and selected one. The following actions are then performed.

- The *fieldList* is cleared.

- The *recordset* pointer is moved to the record in the *gSelection* set that corresponds to the selected value in *cboIDs.*
- For each field in the *gSelection.Fields* collection, if the field is a string, the field name and the field value are added to the *fieldList* list box. If the field is not a string, the field name and the field value as a string are added to the *fieldList* list box.
- The shape that corresponds to the *cboID* flashes three times. This is accomplished with the MO function *FlashShape.*

All you need to do now is to load the form; that is, to figure out what to do when *SelectedFormUp* is true.

6 In the *MouseDown* event (form *Form1*'s code page), find the section for the Identify button. Add the following lines after the *Map1.Refresh* call.

```
If SelectedForm.Visible = True Then
  LoadSelectedForm
Else
  If recs.Count > 0 Then
   LoadSelectedForm
   SelectedForm.Show
  End If
End If
```

Here, if the form is visible, it is updated. If it is not visible, the program checks to see if it qualifies for display on the screen (i.e., if there are any selected records). If it does not qualify, the *LoadSelectedForm* sub is not called.

Now all that remains is to write the *LoadSelectedForm* sub. You will place this form in form *Form1*'s code page as a general, public function.

7 Input the following in form *Form1*'s code page.

```
Public Sub LoadSelectedForm()
Dim curfield as MapObjects2.Field
SelectedFormUp = True
With SelectedForm
  Select Case ActiveLayer.shapeType
  Case moPoint
   .ftypeLabel.Caption = "Point Feature"
  Case moLine
   .ftypeLabel.Caption = "Line Feature"
  Case moPolygon
   .ftypeLabel.Caption = "Polygon Feature"
  End Select
```

```
    If .cboIDs.listCount > 0 Then
     .cboIDs.Clear
    End If
    If .fieldList.listCount > 0 Then
     .fieldList.Clear
    End If
    If gSelection.Count <= 0 Then
     .Numfound.Caption = "No features found"
    Else
     If gSelection.Count = 1 Then
      .Numfound.Caption = "1 feature found"
     Else
      .Numfound.Caption = Str(gSelection.Count) + " features found"
     End If
     gSelection.MoveFirst
     Do While Not gSelection.EOF
       .cboIDs.AddItem gSelection("FeatureID").ValueAsString
      gSelection.MoveNext
     Loop
     .cboIDs.ListIndex = 0
     gSelection.MoveFirst
     Do While Not gSelection.EOF
       .cboIDs.AddItem gSelection("FeatureID").ValueAsString
      gSelection.MoveNext
     Loop
     .cboIDs.ListIndex = 0
     gSelection.MoveFirst
     For Each curfield In gSelection.Fields
      If curfield.Type = moString Then
       .fieldList.AddItem curfield.Name + " = " + _ curfield.Value
      Else
       .fieldList.AddItem curfield.Name + " = " + _ curfield.ValueAsString
      End If
     Next curfield
     Map1.FlashShape gSelection("shape").Value, 3
    End If
  End With
  SelectedFormUp = False
End Sub
```

Let's examine this sub more closely. It begins by declaring a variable for storing the current field (*curfield*). This is used in a *For* loop later in the sub. The Boolean *SelectedFormUp* is set to True. Using a *With* block, the content of *SelectedForm* is placed in the appropriate controls. The *ftypeLabel* caption is set to the type of feature (point, line, or polygon).

The sub then checks to see if anything is in the combo box and the field list. These are cleared out if they are not already empty. The statement *if gSelection.Count <= 0* is needed to check for cases in which *SelectedForm* is displayed and the user uses the Identify button again but selects no features. If there are features, the *Numfound* caption is set to the number of features in the *Else* condition.

The program then populates the *cboIDs* combo box with feature IDs of all records in *gSelection*. This is accomplished in the *Do While Not gSelection.EOF* loop. Once the *cboIDs* combo box is populated, the record pointer in *gSelection* is positioned to the first record (*gSelection.MoveFirst*) and a *For* loop (*For Each curfield In gSelection.Fields*) is used to populate the *fieldList* control with the field name and value for each field for the first selected item.

Summary

In this chapter you have worked with several new properties, methods, events, and objects. The program is starting to have some real power. In particular, you have learned how to create a new selected set and have seen how to draw features with a particular symbol property (the color). In subsequent chapters you will build on these abilities, beginning with rendering features (thematic mapping). In Chapter 12, you will develop more advanced set selection methods.

Chapter 8

Rendering, Part 1: Single Symbols

Introduction

In the last chapter you drew selected features in a special color to differentiate them from features that were not selected. In this chapter and the next three, you will work on building thematic mapping tools. Here, a new VB control, called the tab strip, plays an important role. You will build a new project to learn how to work with this control. You will then take that knowledge and apply it to a legend editor.

The Tab Strip Control

The Tab Strip control is part of the Windows Common Controls. It is added to a form like any other control (see figure 8-1). To see how this feature works, begin with the following steps.

CD-ROM NOTE: *The VB project in the* Chapter8_1 *directory on the companion CD-ROM starts here.*

1. Start a new project. Make sure the Windows Common Control tools are in the toolbar. If they are not, add them. (See Chapter 5 if you do not remember how to do this.)

When you add a tab strip to a form, it is named *TabStrip1* (if it is the first tab strip on the form). You can set some properties for the tab strip by right clicking on the Tab Strip control and selecting Properties (figure 8-2).

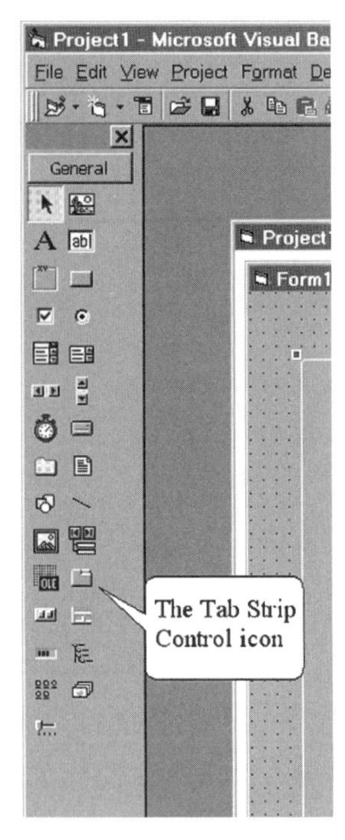

Fig. 8-1. Tab Strip control.

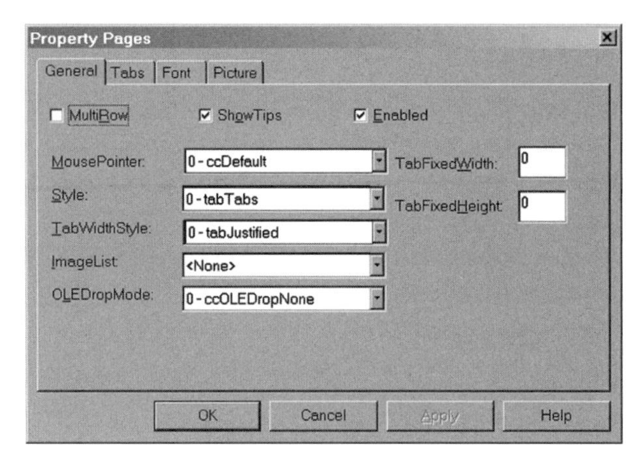

Fig. 8-2. The Tab Strip control's Properties dialog.

Study the General Properties of the Tab Strip control. The Style property can be set to tabs or buttons, and the Image List can be used to place an icon on each tab. Of course, you would have to have an Image List control on the form first. The Tabs tab allows you to add tabs, in much the same way you added buttons to a toolbar in Chapter 6 (see figure 8-3).

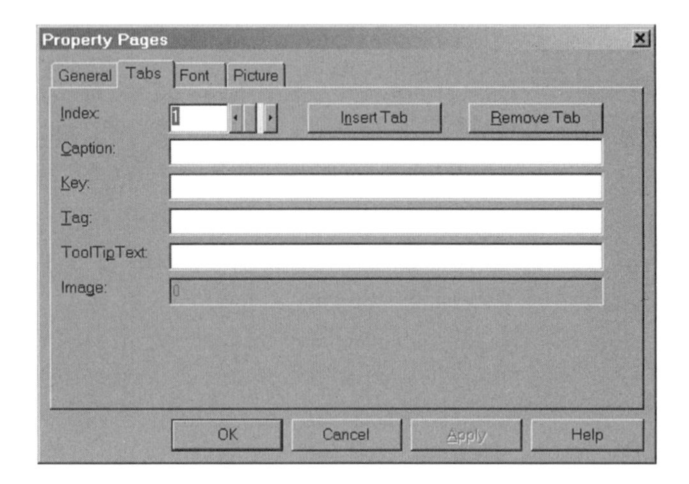

Fig. 8-3. Adding tabs to the Tab Strip control.

If you think about it, this property sheet uses the very control you are trying to learn! When you click on the Tabs tab, it appears that it comes forward, and the page for which it is a tab comes to the front as well. This is the way things appear, but it is *not* the way they really operate. Let's add a few tabs to this control. Figure 8-4 shows what appears when you run the program.

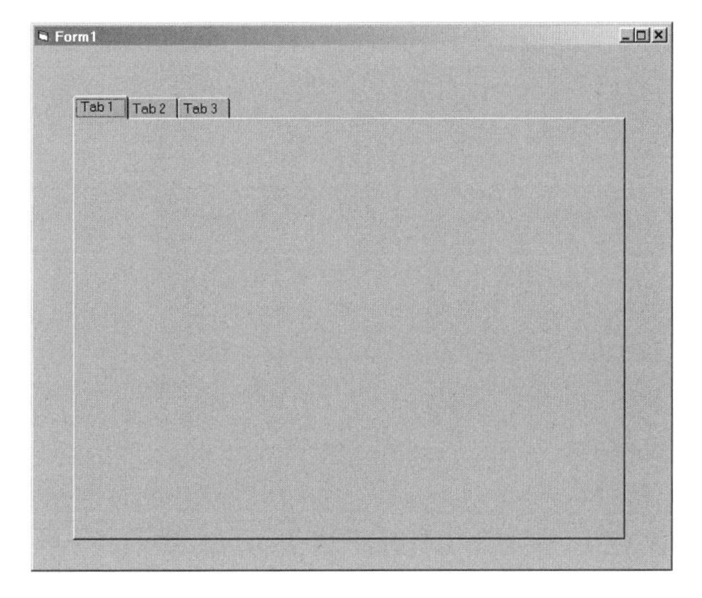

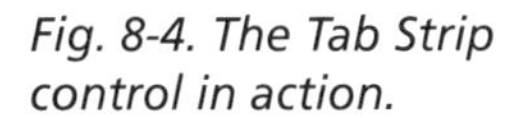
Fig. 8-4. The Tab Strip control in action.

If you click on each tab during execution, the top line changes so that the tab you clicked on moves to the front. In the design form (in the VB IDE), if you click on the Tab Strip control, the page never seems to change, and the tabs do not change order. How do you get different content on each page?

The key is that *there are no pages.* That is, the Tab Strip control is not a *container.* What you need to do is create a series of frames, each frame of which contains the controls for a "virtual page." When the user clicks on a tab, one frame should be turned on, and the other frames off. Think back to the traffic light program of Chapter 1. That involved three pictures of traffic lights, with each picture showing a different light turned on. You changed the traffic light by determining which picture was on top. You will use a similar strategy here.

2 Set the properties of your tab strip so that it has at least two tabs. Place a frame on the form, inside the Tab Strip control. If you want, you can place a few controls (say, buttons and lists) in the frame.

3 Make the frame about one-third the size of the Tab Strip control. Give it a caption of *Frame 1*, and name the frame *TFrame.*

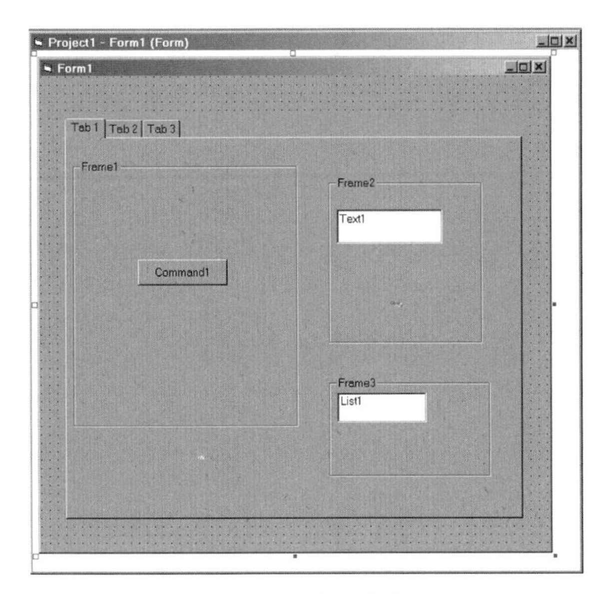

Fig. 8-5. A Tab Strip control with frames.

4 Place another frame on the form inside the Tab Strip control. Again, you can place some controls on this frame (do not make it look exactly like the first frame).

5 Give the second frame a caption of *Frame 2,* and name this frame *TFrame.* Because the frame name is the same as the first frame VB will ask if you really want to do this. Respond in the affirmative.

6 Place as many frames on the form as you have tabs in the Tab Strip control. Figure 8-5 shows a form containing three frames and three tabs.

7 Double click on the *form* to bring up its code page, and add the following code.

```
Option Explicit
Private Sub Form_Load()
  Dim i As Integer
  TFrame(0).Visible = True
  For i = 2 to TabStrip1.Tabs.Count
      TFrame(i-1).Visible = False
  Next
End Sub
Private Sub TabStrip1_Click()
  Dim i As Integer
  For i = 1 to TabStrip1.Tabs.Count
      If TabStrip1.SelectedItem.Index = i Then
           TFrame(i-1).Visible = True
      Else
           TFrame(i-1).Visible = False
      End If
  Next
End Sub
```

The *Form_Load* function establishes the first frame, *TFrame(0)*, as visible, and all other frames as invisible. The *TabStrip1_Click* function is where the synchronization between the frames and the tabs takes place. The first thing to notice is that the tabs are numbered starting with 1, not 0. The frames are numbered starting with 0, not 1. This is why the frames are referenced by subscripts *i-1*, not *i*. If you try this program, you will see that as you click on the tabs the proper frames appear.

You will often want frames to be the same size and overlay each other. This makes for a nice user interface. However, editing overlapping frames can be a bit difficult, particularly if you turn off the border on the frames. Some simple editing tricks to remember are that you can choose the control you edit by clicking on the drop-down list in the Properties box (the middle box on the right side of figure 8-5).

You can also right click on a frame and select Send to Back to see the frame below it. However, you must be careful. Before compiling or executing your program, you *must* make sure that every frame on the tab strip form is above the tab strip form. That is, the tab strip form must be placed behind all frames.

Setting Drawing Properties

CD-ROM NOTE: *The VB project in the* Chapter8_2 *directory on the companion CD-ROM starts here.*

How will you use this new ability in a mapping program? Right now, when you add a layer to your map, MO assigns it a default color, symbol style, and symbol size. Suppose, however, that you wanted to change these default settings. If, for example, MO were to color water bodies red, you would probably want to change this to some shade of blue. In this section, you will develop a frame for changing the symbol for an entire layer. In later chapters you will add other tabs and frames for other types of rendering.

Return to the mapping program you were working on at the end of Chapter 7 (or load the *Chapter8_2* VB project located on the companion CD-ROM). Perform the following steps.

1. Place a new Command button on form *Form2*.
2. Name this button *cmdProps*, and give it a caption of *Properties* (see figure 8-6).

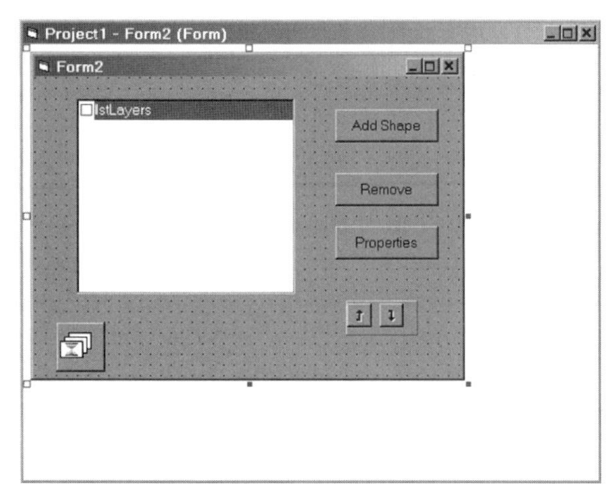

Fig. 8-6. Properties command button.

When should this button be enabled? Only when there is a highlighted item in the *lstLayers* list box; that is, only when *lstLayers.ListIndex > = 0.* However, the way the program is written, there will be a selected item as long as there are items in the list (layers on the map).

3 To provide this functionality, add the following to the *refresh_buttons()* sub in form *Form2*'s code page.

```
If lstLayers.Count > 0 Then
cmdProps.Enabled = True
Else
cmdProps.Enabled = False
```

Another way to do this is to set *cmdProps.Enabled* to True by default. Then, if *lstLayers.Count = 0,* set it to False. You also need a global variable for storing the layer whose properties you want to change.

4 To supply the global variable, in the module *modUtility.bas,* add the following line.

```
Public drawLayer as MapObjects2.Layer
```

When the user clicks on the Properties button, a form should appear, containing a Tab Strip control. Initially, the tab strip will contain just one tab, but you will add more in the coming chapters.

5 Create a form that looks like that shown in figure 8-7. Name it *frmDrawProps.* Incorporate the following controls in this form.

- ❐ A frame that will contain other controls. Name the frame *TFrame.* (Note: In figure 8-7, the frame border style is set to None so that the frame is not visible.)
- ❐ A label box named *lblLayerName.* This will store the name of the layer whose drawing properties you are setting.
- ❐ A picture box control named *pctColor.* This will store a color picture of the current color.

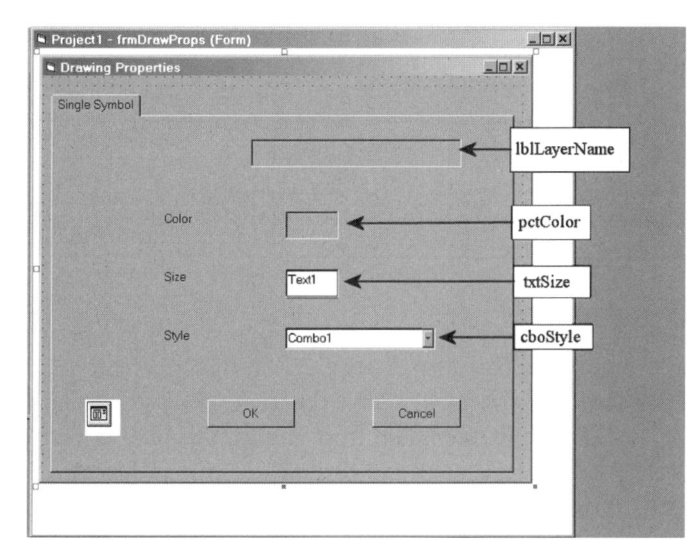

Fig. 8-7. Drawing properties form.

- A textbox called *txtSize.* This will store the size of the symbol (if appropriate).
- A combo box named *cboStyle.* This will allow the user to choose a symbol style.
- Command buttons for OK and Cancel.
- A Common Dialog control, which will be used to display available colors.
- Labels for Color, Size, and Style.

When the form first comes up, the proper values need to be loaded into its controls.

6 Enter the following *Form_Load* code in *frmDrawProps*' code page.

```
Private Sub Form_Load()
  With drawLayer
      lblLayerName = .Name
      pctColor.BackColor = .Symbol.Color
      txtSize = .Symbol.Size
      Select Case .shapeType
      Case moPoint
          cboStyle.AddItem "Circle"
          cboStyle.AddItem "Square"
          cboStyle.AddItem "Triangle"
          cboStyle.AddItem "Cross"
      Case moLine
          cboStyle.AddItem "Solid Line"
          cboStyle.AddItem "Dash Line"
          cboStyle.AddItem "Dot Line"
          cboStyle.AddItem "Dash Dot"
          cboStyle.AddItem "Dash Dot Dot"
      Case moPolygon
          Label2.Visible = False
          txtSize.Visible = False
          cboStyle.AddItem "Solid Fill"
          cboStyle.AddItem "Transparent"
```

```
      cboStyle.AddItem "Horizontal"
      cboStyle.AddItem "Verical"
      cboStyle.AddItem "Upward Diagonal"
      cboStyle.AddItem "Downward Diagonal"
      cboStyle.AddItem "Cross"
      cboStyle.AddItem "Diagonal Cross"
    End Select
    cboStyle.ListIndex = .Symbol.Style
  End With
End Sub
```

With this code, when the form loads, the name of the *drawLayer* is loaded into the label box *lblLayerName*, the color in *pctColor.BackColor* is set to *drawLayer.Symbol.Color*, and the *txtSize* box is set to *drawLayer.Symbol.Size.* Note that if the draw layer is a polygon, the label and size fields are made invisible (*Visible = False*). This is done because a size field has meaning for lines and points only.

Next, the types of symbols in the combo box are initialized. In MO, each shape type (point, line, polygon) has a set of default symbols that can be used. Each symbol type has an integer value associated with it, and these values (for a given shape type) start from 0. This is very helpful. Once the content of the combo box list is built, all you need to know is the *ListIndex* value. This is because the *ListIndex* value will be the same as the draw symbol's value. How do you find the draw symbol's value? Examine figure 8-8.

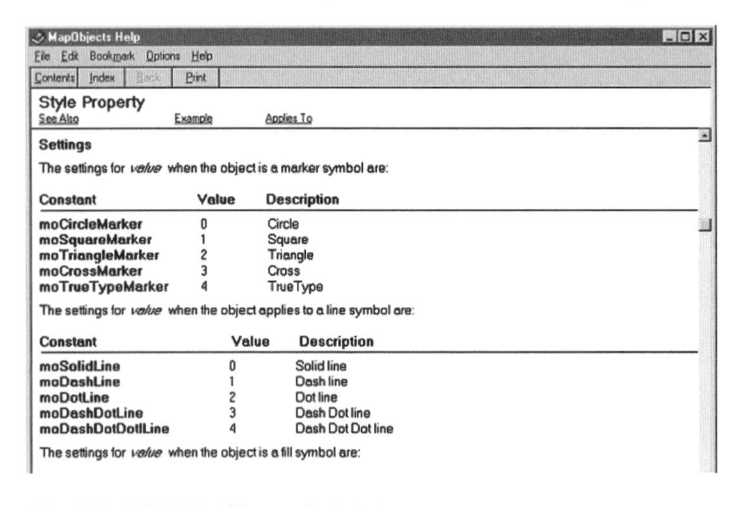

MapObjects Help

File Edit Bookmark Options Help

Contents Index Back Print

Style Property

See Also Example Applies To

Settings

The settings for *value* when the object is a marker symbol are:

Constant	Value	Description
moCircleMarker	0	Circle
moSquareMarker	1	Square
moTriangleMarker	2	Triangle
moCrossMarker	3	Cross
moTrueTypeMarker	4	TrueType

The settings for *value* when the object applies to a line symbol are:

Constant	Value	Description
moSolidLine	0	Solid line
moDashLine	1	Dash line
moDotLine	2	Dot line
moDashDotLine	3	Dash Dot line
moDashDotDotlLine	4	Dash Dot Dot line

The settings for *value* when the object is a fill symbol are:

Fig. 8-8. MO Help file symbol styles.

The last line of the *Form_Load* function sets *ListIndex* equal to the current draw symbol. Because this is a number, it sets the proper item in the combo box *cbo.Style.* The form will function as shown in figure 8-9.

In the example depicted in figure 8-9, the current draw symbol is a square, which has the value of 1. If you look at *Form_Load,* you will see that the square is the second item in the combo box list. That is, it has an index value of 1.

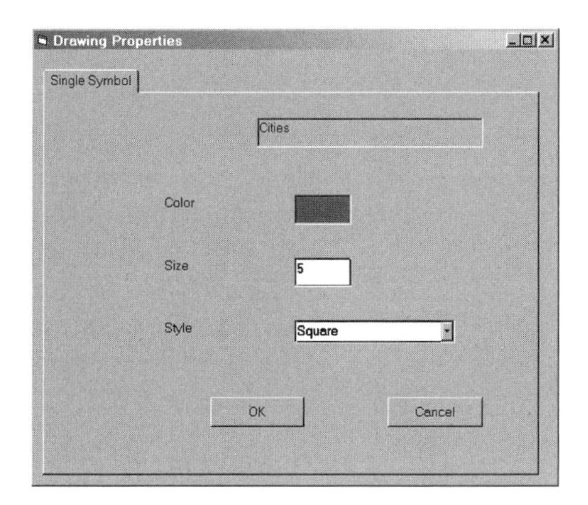

Now suppose someone were to click on the color picture box. What should happen? Here is where you will use the common dialog box for colors.

7 Enter the following sub in *frmDrawProps'* code page.

Fig. 8-9. Single-symbol Drawing Properties dialog.

```
Private Sub pctColor_Click()
  Dim curcolor As Long
  CommonDialog1.CancelError = True
  On Error GoTo ErrHandler
  CommonDialog1.ShowColor
      curcolor = CommonDialog1.Color
      pctColor.BackColor = curcolor
  Exit Sub
ErrHandler:
  'do nothing, just exit
      Exit Sub
End Sub
```

In this code, a local variable, *curcolor,* is dimensioned to store the color choice. This is a long integer. The Common dialog's *ShowColor* method is used to display a palette of available colors (figure 8-10). *CancelError* is set to True to capture cases in which the user selects Cancel (does not want to select a new color). VB is then told that if an error occurs (the user selects Cancel) the program should jump to the *ErrHandler* portion of the sub.

If the user clicks on OK, there is no error and the program should set *curcolor* to the value the user chooses, reset the *pctColor* box to that color, and exit the sub. If the user does select Cancel, at this point the program should simply exit the *pctColor_Click* sub.

If the user changes the size of the draw symbol, the program simply records the change. If you wanted to be more careful, you could incorporate an error-checking function in the code that would make sure the new size value was an integer. You could also use a combo box of sizes rather than a text box. This is probably a better alternative; however, you have to make sure that the combo box list contains the current value. Finally, you need to manage changes to the style combo box.

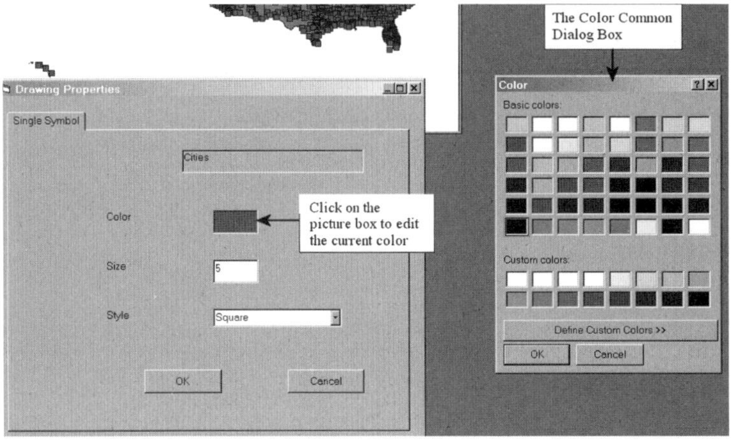

Fig. 8-10. Common dialog color chart.

Now all controls in the frame are accounted for except the Cancel and OK buttons on *frmDrawProps*. If the user were to click on Cancel, the form should close.

8 Enter the following code in *frmDrawProps*' code page.

```
Private Sub cmdCancel_Click()
Unload Me
End Sub
```

If the user were to click on OK, the *drawLayer*'s symbol object should be reset with its new properties (those in the dialog box).

9 Enter the following code in *frmDrawProps*' code page.

```
Private Sub cmdOK_Click()
With drawLayer.Symbol
    .Color = pctColor.BackColor
    .Style = cboStyle.ListIndex
    If drawLayer.shapeType <> moPolygon Then
        .Size = txtSize
    End If
End With
Form1.Map1.Refresh
Unload Me
End Sub
```

Summary

This completes the first tab of the drawing properties. You learned how to set the symbol type, color, and size for an entire layer. In the next chapter, you will continue to build on your knowledge of layer controls, allowing for thematic mapping of layers.

Chapter 9

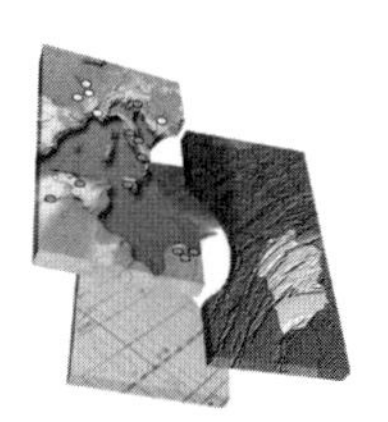

Rendering, Part 2: The Unique-value Map Renderer

■■ Introduction

This chapter and the next explore a type of thematic map renderer—the unique-value map. In Chapter 8, you saw how to set a symbol for an entire layer, but what if you want to map features based on their values in a database? Conceptually, this is a straightforward process. Operationally, however, there is a bit of overhead.

CD-ROM NOTE: *The VB project in the* Chapter9_1 *directory on the companion CD-ROM starts here.*

At the end of the last chapter, you worked with a drawing properties form (*frmDrawProperties*), which looked like the form shown in figure 9-1. You will need to change this form a bit. Open the VB project so that you can edit the *frmDrawProperties* form.

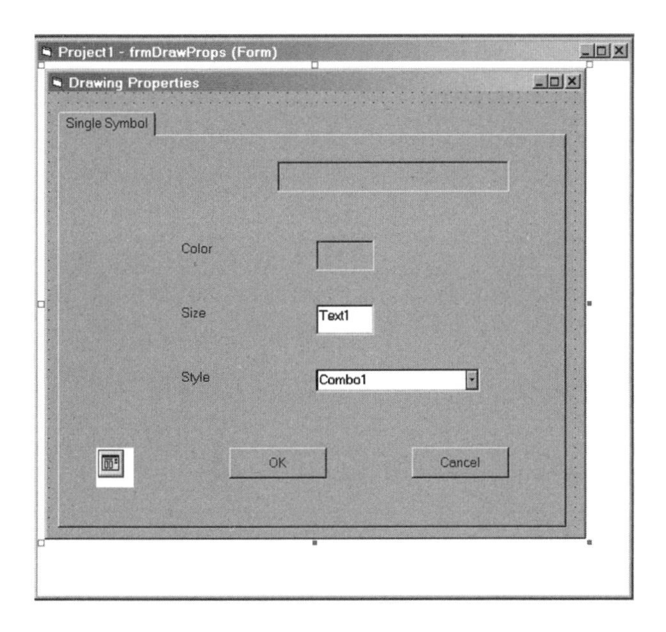

Fig. 9-1. The frmDrawProperties *form.*

1 Enlarge the form box so that there is room above and below the Tab Strip control for some other controls.

Select the Common Dialog control, the OK button, and the Cancel button. *Cut* these off the tab strip and *paste* them below the form.

2 Cut the label box (*lblLayerName*) and paste it above the form.

3 Add a new tab and give it the caption *Unique Value.*

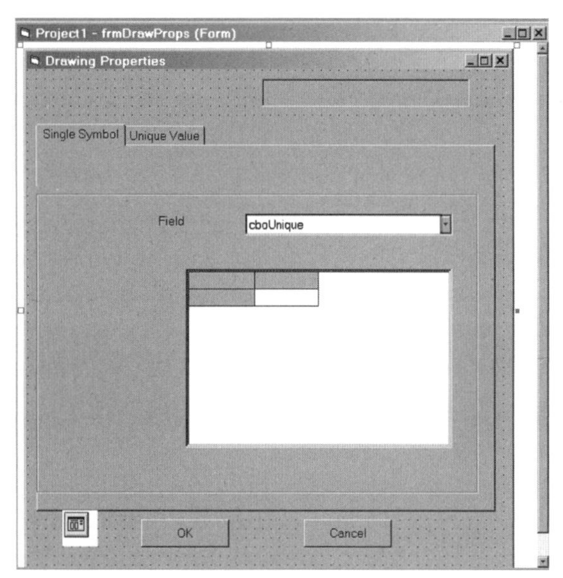

The form should now look like that shown in figure 9-2.

By moving these controls out of the first frame (*TFrame*) and onto the form, they now become available to all frames (which will be associated with tabs) you create.

Fig. 9-2. New arrangement of the frmDrawProperties *form.*

4 Create a new map control on this form. Name this control *mapDrawSymbol.* Set its height to 310 and its width to 864.

For now, you might want to leave the control where it is visible on the form—just to see how things work. However, after you finish the renderer, you will want to hide the map control behind the Tab Strip control so that the user does not see it.

NOTE: Do not place the map control into a frame.

If you want to hide the map control, drag the control so that it is "behind" the Tab Strip control. Be sure to drag it, not to cut it and paste it. This will hide the map window behind the Tab Strip control. If it still is visible, right click on it and then select Send To Back.

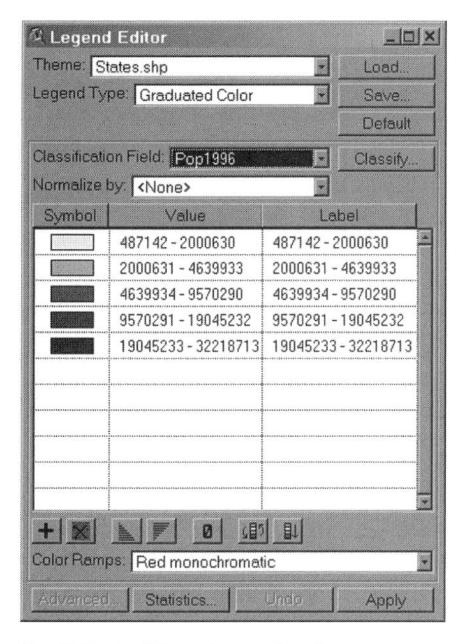

Fig. 9-3. ArcView legend editor.

Why would you want a map window the user would never see? You will use this map control in a very clever way. For each drawing possibility, you will draw the symbol in the map control, and then paste the resulting "map" in the legend editor. The resulting legend editor will look something like that shown in figure 9-3 (from ArcView 3).

In order to draw each "symbol map," you will use a class (classes are discussed in material to follow and in more detail in later chapters). The class you will use is based on a class from *MoView2*. However, a few modifications have been made.

CD-ROM NOTE: *The resulting class is contained on the companion CD-ROM in a file named* DrawSymbol.cls. *This file contains code for creating a new class.*

Perform the following.

5. Copy the file *DrawSymbol.cls* to your project directory.
6. Add this file to your project. To do this, click on Project, click on Add Class Module, and then select the tab named Existing and open the file you just copied.

▪▪ VB and Classes

In VB, a class module contains the code for a user-defined object. As an object, it will have properties and methods. Let's take a look at the class you just copied and added to the project. The class starts with the following declarations.

```
Dim curMap As Object
Dim cursymbol As MapObjects2.symbol
```

curMap is the name of an object (just what type of object you do not know at this point). The class also has a property called *cursymbol* (a MapObjects symbol object). The class has the following *Get* and *Set* properties.

```
Public Property Get symbol()
End Property
Public Property Set symbol(symbol As MapObjects2.symbol)
 'This procedure sets the symbol and draws it to the map control.
 Set cursymbol = symbol
End Property
```

The first two functions are the *Get* and *Set* procedures for *symbol.* Let's look at the second method, *Set symbol.* In this method, the user passes an MO symbol object, and the object's *cursymbol* is set to that symbol.

NOTE: *When you use* Set, *you do not make a copy of symbol, you simply create a reference. That is,* cursymbol *points to the memory where symbol is located. You might want to review the discussions of "pass-by-value" and "pass-by-reference" in Chapter 3.*

Get contains no value, and therefore the *cursymbol* property is write-only (i.e., you can set it, but you cannot get it). The same is true of the *curmap* object, as can be seen from its *Get* and *Set* properties in the following.

```
Public Property Get mapControl()
End Property
Public Property Set mapControl(curControl As Control)
 Set curMap = curControl
End Property
```

The next method uses these properties to draw *cursymbol* in *curmap,* as follows.

```
Public Sub Draw()
 'This procedure draws the settings of the current symbol onto
 'the map control on the form. A single instance of the symbol
 'is drawn in the center of the map.
 'These geometric objects are used for drawing symbols
 Dim drawLine As New MapObjects2.Line
 Dim drawPoint As New MapObjects2.Point
 Dim drawPoints As New MapObjects2.Points
 Dim drawRect As New MapObjects2.Rectangle
```

The next section of code is different from ESRI's original class. The following code creates a rectangle object with the same dimensions as the map control. It is these dimensions you will use to create shapes for drawing in *mapDrawSymbol.* Those shapes will then be copied to the legend editor.

```
Dim currect As New MapObjects2.Rectangle
Set currect = curMap.Extent
```

```
Select Case frmDrawProps.curFeatureType
Case moPoint
   'Just make a point and draw it in the center of the control
   drawPoint.X = currect.Left + currect.Width / 2
   drawPoint.Y = currect.Height / 2
   curMap.DrawShape drawPoint, cursymbol
Case moLine
   'Add the end points to a points collection
   drawPoint.X = currect.Left
   drawPoint.Y = currect.Height / 2
   drawPoints.Add drawPoint
   drawPoint.X = currect.Left + currect.Width
   drawPoint.Y = currect.Height / 2
   drawPoints.Add drawPoint
   drawLine.Parts.Add drawPoints
   curMap.DrawShape drawLine, cursymbol
Case moPolygon
   'Size the rectangle to be a little smaller than the map.
   Dim margin As Integer
   drawRect.Top = currect.Top
   drawRect.Left = currect.Left
   drawRect.Bottom = currect.Bottom
   drawRect.Right = currect.Right
      curMap.DrawShape drawRect, cursymbol
 End Select
End Sub
```

In the next section you will use this class as follows. You will create an instance of the class, with *curmap* being the map control you added to the form *frmDrawProps* at the beginning of this chapter. You will then pass the instance of a symbol and then call the *Draw* method. The current symbol will be drawn in the map control. You will copy that map to the Windows clipboard, and then paste it into the legend.

Editing the frmDrawProps Code Page

Because you will be adding new elements to the tab strip, you need to adjust the existing code.

1. At the top of the *frmDrawProps* code page, add the following declarations.

```
Option Explicit
Public curTab As Integer
```

```
Public vmr As New MapObjects2.ValueMapRenderer
Public tempSymbol As New MapObjects2.symbol
Public curDrawSymbol As New clsDrawSymbol 'Note the use of the class
Public curFeatureType As Integer
Public curFeatureName As String
Dim tabUp As Boolean
```

2 Edit *formDrawProps' Form_Load* function to read as follows.

```
Private Sub Form_Load()
   Dim i As Integer
   lblLayerName = drawLayer.Name
   'When first loaded into the map, all layers have a blank tag
   If drawLayer.Tag = "" Then drawLayer.Tag = "SingleSymbol"
   Select Case drawLayer.Tag
   Case "SingleSymbol"
      curTab = 1
      RestoreSingleValueMap
   Case "UniqueValue"
      curTab = 2
      RestoreUniqueValueMap
   End Select
   For i = 1 To TabStrip1.Tabs.Count
   If curTab = i Then
         TFrame(i - 1).Visible = True
      Else
         TFrame(i - 1).Visible = False
      End If
   Next
End Sub
```

This code starts by setting the value in *lblLayerName* to the *drawLayer*'s name. It then checks to see if the *Tag* property of the layer is set. The tag property is a "special" property you can set to whatever you want. You will use this property to store the renderer type for each layer. When layers are first entered into the project, their tag is blank, which you will treat the same as *SingleSymbol.* A select on *Tag* is performed. This will display the frame corresponding to how the layer is currently drawn.

Assume that *drawLayer* is a single-symbol map. The select case *drawLayer.Tag* will go to the *SingleSymbol* case. This will result in a call to *RestoreSingleValueMap.*

3 The content of the *RestoreSingleValueMap* sub is what used to reside in *Form_Load.* In *frmDrawProps'* code page, add the following sub.

```
Private Sub RestoreSingleValueMap()
  With drawLayer
     pctColor.BackColor = .symbol.Color
     txtSize = .symbol.Size
     cboStyle.Clear
     Select Case .shapeType
     Case moPoint
        cboStyle.AddItem "Circle"
        cboStyle.AddItem "Square"
        cboStyle.AddItem "Triangle"
        cboStyle.AddItem "Cross"
     Case moLine
        cboStyle.AddItem "Solid Line"
        cboStyle.AddItem "Dash Line"
        cboStyle.AddItem "Dot Line"
        cboStyle.AddItem "Dash Dot"
        cboStyle.AddItem "Dash Dot Dot"
     Case moPolygon
        Label2.Visible = False
        txtSize.Visible = False
        cboStyle.AddItem "Solid Fill"
        cboStyle.AddItem "Transparent"
        cboStyle.AddItem "Horizontal"
        cboStyle.AddItem "Verical"
        cboStyle.AddItem "Upward Diagonal"
        cboStyle.AddItem "Downward Diagonal"
        cboStyle.AddItem "Cross"
        cboStyle.AddItem "Diagonal Cross"
     End Select
     cboStyle.ListIndex = .symbol.Style
  End With
End Sub
```

This sub loads the single-symbol tab and its corresponding frame with appropriate symbol type, color, and size. In the previous chapter, the OK button corresponded only to *frmDrawProps*' *Single-Symbol* tab. However, there are now two tabs in *frmDrawProps*, and therefore the OK button has to be modified to handle both cases.

4 Edit *frmDrawProps*' *cmdOK_Click* sub to read as follows.

```
Private Sub cmdOK_Click()
  Select Case TabStrip1.SelectedItem.index
  Case 1
    Set drawLayer.Renderer = Nothing
    drawLayer.Tag = ""
    With drawLayer.symbol
```

```
            .Color = pctColor.BackColor
            .Style = cboStyle.ListIndex
            If drawLayer.shapeType <> moPolygon Then
               .Size = txtSize
            End If
         End With
      Case 2
         'We will develop this below
      End Select
      Form1.Map1.Refresh
      Unload Me
   End Sub
```

You now have created a second tab on *frmDrawProps* for the unique-value map renderer. Adding this second tab required several changes to the *frmDrawProps* form and its code. A new Map control was added to the form, the *Form_Load* function was rewritten to handle two possible renderers, the code for initializing the single-symbol case was moved from the *Form_Load* sub to the *RestoreSingle-ValueMap* sub, and the *cmdOK_Click* sub was edited to handle multiple renderers. These changes were necessary to accommodate the unique-value map renderer. It is now time to develop that renderer.

Creating the Value Map Renderer

MO has several types of renderers: the unique-value map renderer (each unique value is assigned its own symbol), the dot density map renderer (which makes sense only for polygons), the class breaks renderer (which can be based on several statistical methods of classification), and the label renderer (for text). Each layer can have no renderer (just a single symbol for the entire layer) or one—and only one (excluding labels)—of these renderers. Let's first explore the value map renderer.

The value map renderer object has several properties. Two you will use in this chapter are the *Field* property and the *Value* property. The *Field* property stores the field upon which the renderer is based. The *Value* property is a collection of unique values found in the field.

1. Start by adding the Microsoft Flex Grid control to your project (figure 9-4).

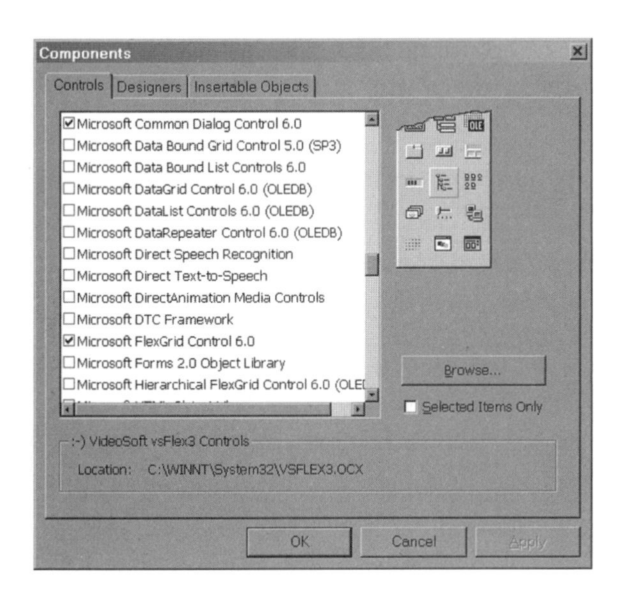

Fig. 9-4. Adding the Flex Grid control.

The Flex Grid control places a grid on a form. This is a very powerful control, particularly for data analysis by such things as pivot tables. You will use only a portion of its power. In particular, you will use its power to hold pictures and text in a cell, as well as its ability to resize and select grid cells. There is a lot more it can do. For example, flex grids have find-and-replace capabilities, sorting capabilities, column resize capabilities, and so on. As you might have guessed, much of the power you find in spreadsheet programs lies in the Flex Grid control. For more details on the Flex Grid's methods and properties, see the VB help.

2 Create a new frame on *frmDrawProps*' Tab Strip control, and name it *TFrame*.

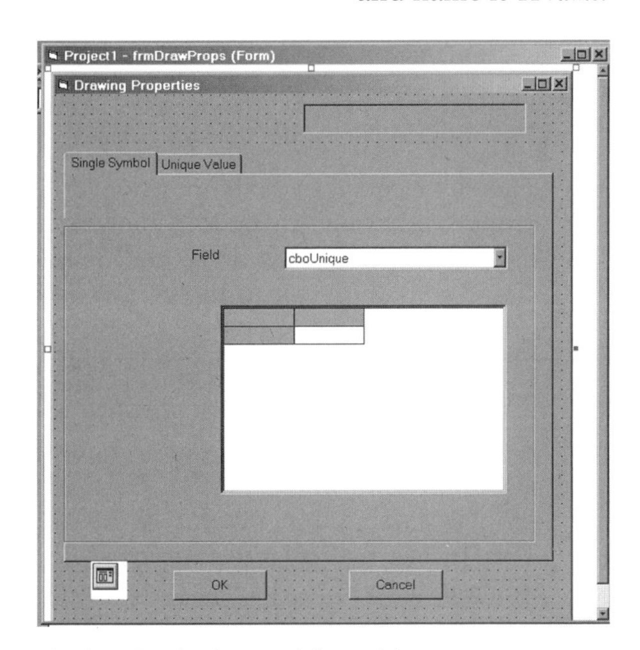

Fig. 9-5. Combo box and flex grid.

This will create an array of frames—one for each tab on the tab strip.

3 In *TFrame(1)*–the new frame–add the following:

- A combo box named *cboUnique*. Set its style to *drop-down list*.
- A flex grid named *grdValues*. Set the highlight to *flexHighlightNever*.

Your form should now look like that shown in figure 9-5.

These new controls will be incorporated as follows. When the user clicks on the Unique Value tab, the *cboUnique* box will be loaded with a list of data fields on which the renderer can be based. When the user chooses a variable from the combo box, the flex grid will be loaded with the unique values for that variable and the corresponding drawing symbol for each unique value.

Let's begin by specifying the *RestoreUniqueValueMap* sub that is called from the *Form_Load* sub.

4 Enter the following.

```
Private Sub RestoreUniqueValueMap()
  Dim recs As New MapObjects2.Recordset
  Set recs = drawLayer.Records
  Dim fld As MapObjects2.Field
  If drawLayer.Tag <> "UniqueValue" Then
     cboUnique.Clear
     cboUnique.AddItem "None"
     cboUnique.ListIndex = 0
     'the layer is not currently a unique value map
        For Each fld In recs.Fields
           If fld.Name <> "Shape" Then
             If fld.Name <> "FeatureId" Then
              cboUnique.AddItem fld.Name
           End If
           End If
        Next
     cboUnique.ListIndex = 0
  'You will develop an "else case" in the next chapter
  End If
```

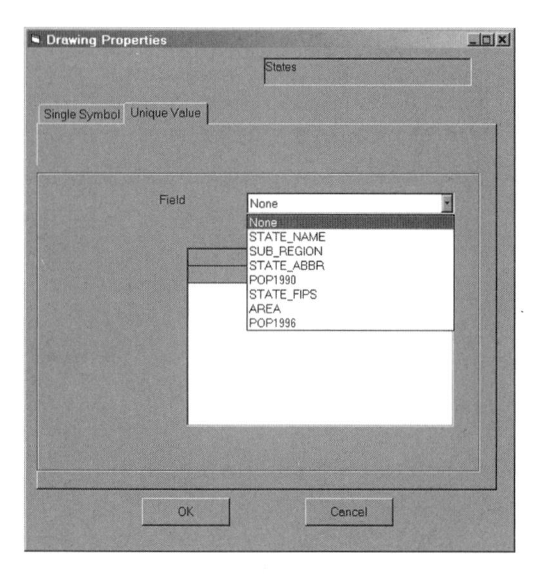

This sub clears the combo box, and adds a value of "None" as the first element in the box. It then adds every field in the *drawLayers* record set to the combo box, except the *Shape* and *FeatureId* fields. Figure 9-6 shows how this sub would look in action.

The key to creating the renderer is deciding what to do when the user chooses a variable (a selection other than None). This is handled by the *cboUnique_Click* sub, as follows.

Fig. 9-6. Combo box with fields added.

5 Enter the following.

```
Private Sub cboUnique_Click()
   Dim recs As New MapObjects2.Recordset
   Dim uniquevals As New MapObjects2.Strings
   Dim retval As Integer
   Dim goOn As Boolean
   Dim i As Integer
   If tabUp Then
      Exit Sub
   End If
   Const tpi = 1440 'Twips per inch
   'Size the grid initially to 2 by 2 and grow as needed.
   grdValues.Clear
   grdValues.Cols = 2
   grdValues.Rows = 2
   'Position on first row and set first column settings
   grdValues.Row = 0: grdValues.Col = 0
   grdValues.ColWidth(0) = tpi * 1.3
   'These set the size of the boxes. You can change these.
   grdValues.ColAlignment(0) = 1 'right align
   grdValues.Text = "Value"
   'Second column settings
   grdValues.Col = 1
   grdValues.ColWidth(1) = tpi * 0.6
   grdValues.Text = "Symbol"
   grdValues.FixedRows = 1: grdValues.FixedCols = 0
      If cboUnique.List(cboUnique.ListIndex) <> "None" Then
      'Load up the grid
       Set recs = drawLayer.Records
       recs.MoveFirst
       goOn = False
       Do While Not recs.EOF
          uniquevals.Add recs(cboUnique.Text).Value
          If uniquevals.Count > 25 And goOn = False Then
             retval = MsgBox("There are more than 25 unique values._
                Continue?", vbYesNo)
             If retval = vbNo Then
                Exit Do
             Else
                goOn = True
             End If
          End If
          recs.MoveNext
       Loop
       'Have the list, build the grid
       'Set drawLayer.Renderer = vmr
```

```
      curFeatureType = drawLayer.shapeType
      Select Case drawLayer.shapeType
      Case moPoint
         vmr.SymbolType = moPointSymbol
      Case moLine
         vmr.SymbolType = moLineSymbol
      Case moPolygon
         vmr.SymbolType = moFillSymbol
      End Select
      vmr.ValueCount = uniquevals.Count
      vmr.Field = cboUnique.Text
      'Add the values and pictures to the flex grid grdValues
      For i = 0 To vmr.ValueCount - 1
         grdValues.Row = i + 1
         grdValues.Col = 0
         grdValues.Text = uniquevals(i)
         'Go to second column
         grdValues.Col = 1
         vmr.Value(i) = uniquevals(i)
         Set tempSymbol = vmr.symbol(i)
         MapDrawSymbol.TrackingLayer.Refresh True
         MapDrawSymbol.CopyMap 1
         Set grdValues.CellPicture = Clipboard.GetData
         grdValues.CellPictureAlignment = flexAlignCenterCenter
         grdValues.Rows = grdValues.Rows + 1
      Next
      'Remove blank line at end
      grdValues.Rows = grdValues.Rows - 1
      grdValues.RowHeight(-1) = tpi * 0.25
  End If
End Sub
```

Let's consider how this sub works. This code starts by determining if the *TabStrip* has just come up (similar to the *FromUp* trick you used elsewhere). If the form is just coming up, the program exits the sub. The program then sets the number of twips per inch. (A twip is a device-independent measure, equal to 1/20 of a printer's point.)

The program then clears the grid, and sets it to two rows and two columns. In the first row and column (*row = 0, col = 0*), the program sets the column width to 1.3 inches. This is arbitrary; you could choose another value. The text alignment is then set to right, and the text in the first row/first column is set to *"Value"*. In the second column (*col = 1*), the width is set to to 0.6 inches, and the text is set to *"Symbol"*. Finally, the first row is set as the header row (always visible, never scrolls off) by setting the fixed rows equal to 1.

The program then needs to check whether or not there is a variable selected. If so, it creates a list (an MO *Strings* object) of unique values. (If you check the MO Help file on the strings collection, you will see that the default is to store only unique values.) This is accomplished by the following lines of code.

```
If cboUnique.List(cboUnique.ListIndex) <> "None" Then
  'Load up the grid
     Set recs = drawLayer.Records
     recs.MoveFirst
     goOn = False
     Do While Not recs.EOF
        uniquevals.Add recs(cboUnique.Text).Value
        If uniquevals.Count > 25 And goOn = False Then
           retval = MsgBox("There are more than 25 unique values. _
                Continue?", vbYesNo)
           If retval = vbNo Then
              Exit Do
           Else
              goOn = True
           End If
        End If
        recs.MoveNext
     Loop
```

When this loop is finished, there exists a collection (*Strings*) of all unique values. Note that the user is asked if she wants more than 25 cases, as colors will repeat. (If there are more than 25 unique values, this is probably an inappropriate renderer!)

The next section of the sub sets *curSymbolType* (used in the *DrawSymbol* class) to the current layer shape type. It then begins to set the properties of *vmr*—the value map renderer declared at the top of the module. You will use this to store the symbols on the legend form. Only when the user decides to keep these (clicks on the OK button) are they committed to *drawLayer.*

A *For* loop is then used to populate the flex grid. The first column in each row is set to contain to the unique value's text, and the second column its symbol. The portion of code that accomplishes this follows.

```
For i = 0 To vmr.ValueCount - 1
   grdValues.Row = i + 1
   grdValues.Col = 0
```

```
        grdValues.Text = uniquevals(i)
        'go to second column
        grdValues.Col = 1
        vmr.Value(i) = uniquevals(i)
        Set tempSymbol = vmr.symbol(i)
        mapDrawSymbol.TrackingLayer.Refresh True
        mapDrawSymbol.CopyMap 1
        Set grdValues.CellPicture = Clipboard.GetData
        grdValues.CellPictureAlignment = flexAlignCenterCenter
        grdValues.Rows = grdValues.Rows + 1
    Next
```

This code starts at row 1 (the second row), column 0 and sets the value to the first string (*uniquevals(0)*) in the list. The code then moves to the second column. Each unique value of the field is added to the value map renderer's value array (*vmr.Value(i) = uniqevals(i)*). A reference is then set from *tempSymbol* to *vmr.symbol(i)*. What happens next is tricky.

A call is made to *TrackingLayer.Refresh*, with the value of True. This is an MO method that causes the tracking layer (which is empty) to be redrawn. This empty layer is drawn so that the program can force a call to *AfterTrackingLayerDraw*. (This is discussed in material to follow.) For now, assume that this causes *tempSymbol* to be drawn on the map control *mapDrawSymbol*. Next, a call is issued to *CopyMap*, with a scale factor of 1. This copies the content of the map control to the Windows clipboard. The 1 sets the scale factor at 1.

Every cell in a flex grid can have a *CellPicture* property. In the previous code, it is set to *Clipboard.GetData*. This is a VB call that gets the content of the clipboard and assigns it to the cell picture. The picture is centered in the cell, and a new row is added to the flex grid. If you left the map control, *mapDrawSymbol*, visible (not behind the Tab Strip control), you can watch the symbol maps being drawn and then pasted into the flex grid.

Finally, to complete the *For* loop the last (empty) row in the flex grid is removed, the row height for all rows (–1) is set to 1/4 inch, and the sub ends. The only things remaining to be done are the calls to *AfterTrackingLayerDraw* and *TabStrip1_Click*.

6 Enter the following in *frmDrawProps*' code page.

```
Private Sub mapDrawSymbol_AfterTrackingLayerDraw(ByVal hDC As_
       StdOle.OLE_HANDLE)
 If cboUnique.ListIndex > 0 Then
    Set curDrawSymbol.mapControl = Me.mapDrawSymbol
```

```
    Set curDrawSymbol.symbol = tempSymbol
    curDrawSymbol.Draw
 End If
End Sub
```

This sub checks to see if there are values to be drawn (i.e., the user has not chosen None). If there are values to be drawn, the two properties of the class *DrawSymbol* (*map control* and *symbol*) are set. Next, the class's *Draw* method is called. This will place the symbol in the map control so that it can be copied to the clipboard. The next sub to change is *TabStrip1_Click*.

7 Enter the following.

```
Private Sub TabStrip1_Click()
  Dim i As Integer
  For i = 1 To TabStrip1.Tabs.Count
     If TabStrip1.SelectedItem.index = i Then
        TFrame(i - 1).Visible = True
     Else
        TFrame(i - 1).Visible = False
     End If
  Next
  Select Case TabStrip1.SelectedItem.index
  Case 1
     RestoreSingleValueMap
  Case 2
     tabUp = True
     RestoreUniqueValueMap
     tabUp = False
  End Select
End Sub
```

Here, the proper frame is made visible, and then the appropriate *Restore* function is called. Note that *tabUp* Boolean is set the same way as for *FromUp* in form *Form2*. The last thing you need to handle is how the program behaves when the user clicks on OK.

8 To the *cmdOK_click* sub, add the following case.

```
Case 2
'Set drawLayer.Renderer = vmr
Dim rv As New MapObjects2.ValueMapRenderer
Set drawLayer.Renderer = rv
rv.Field = vmr.Field
rv.SymbolType = vmr.SymbolType
rv.ValueCount = vmr.ValueCount
Dim i As Integer
```

```
For i = 0 To vmr.ValueCount - 1
  rv.Value(i) = vmr.Value(i)
  rv.symbol(i).Color = vmr.symbol(i).Color
  rv.symbol(i).Style = vmr.symbol(i).Style
  rv.symbol(i).Size = vmr.symbol(i).Size
Next
drawLayer.Tag = "UniqueValue"
```

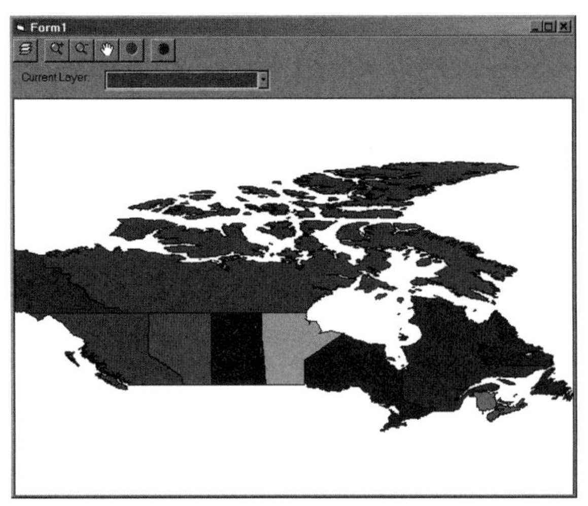

Fig. 9-7. Canadian provinces displayed via a unique-value map renderer.

You might think the first commented line would work—setting the *drawLayer*'s renderer to *vmr.* Well, it will work, but only for a while. This is because *Set* is used, which sets a reference. Recall the discussion of the Pass-by-Value and Pass-by-Reference options in Chapter 3. By using *Set*, the layer's renderer points to a location in memory. If the content of that location changes—for example, if the user sets the renderer for a second layer—the previous renderer will be overwritten!

Suppose you had *Lakes* and *States* layers and set the Lakes renderer. Then you went back and set the States renderer. The Lakes renderer would be overwritten and the lakes would be drawn as an empty outline. This is why you have to create a new *MapObjects2.ValueMapRenderer*, *rv*, and copy over (by value!) all of the necessary properties (*Color*, *Style*, and *Size*). You can now use the program to create unique-value map renderings (see figure 9-7).

■■ Summary

Wow—that was a lot, yet you have not finished this renderer. You need to figure out a few other things. You need to know what to do if the user clicks on a symbol in the flex grid. How will it be changed? You also need to figure out how to initialize the flex grid if the layer is already drawn with a value map renderer. Finally, you could create different flex grids. Instead of having a picture in each row, you could add columns for color, style, and size for each unique value. There are many ways to build a legend editor, and how you choose to do so is somewhat a matter of personal taste.

Chapter 10

The Unique-value Map Renderer Continued

■■ Introduction

In the previous chapter you added a new tab and saw how to initialize a value map renderer. You added a sub called *cboUnique_Click*. You placed the following code in this sub.

```
For i = 0 To vmr.ValueCount - 1
   grdValues.Row = i + 1
   grdValues.Col = 0
   grdValues.Text = uniquevals(i)
   'go to second column
   grdValues.Col = 1
   vmr.Value(i) = uniquevals(i)
   Set tempSymbol = vmr.symbol(i)
   mapDrawSymbol.TrackingLayer.Refresh True
   mapDrawSymbol.CopyMap 1
   Set grdValues.CellPicture = Clipboard.GetData
   grdValues.CellPictureAlignment = flexAlignCenterCenter
   grdValues.Rows = grdValues.Rows + 1
Next
```

This worked just fine, but it is a bit dangerous. Can you see why? Look at where you set *tempSymbol*. You used a *Set* operator (think pass-by-reference). This caused *tempSymbol* to "point" to the memory used by *vmr.symbol(i)*. Now consider what happens if you want to give the user the ability to change a symbol for a particular value in the *vmr.*

You could set the draw symbol equal to the *vmr* symbol, as in the previous code. You could then change the color, for example, of *tempSymbol.* If you do this (and you will in this chapter), you have changed the value of *vmr.symbol(i).* If the user then decides to discard the change (clicks on Cancel), the map will not change, but the *vmr* will change. That is, the color in the Draw Properties form may not match the color on the map. (This will become clearer in material follow.)

CD-ROM NOTE: *The VB project in the* Chapter10_1 *directory on the companion CD-ROM starts here.*

The point is that you must be very careful with references. The following is a safer method. Whenever you use *tempSymbol,* you first initialize it to "nothing." You then copy over, value by value, the relevant properties of *vmr.symbol(i)*: symbol type, size, color, and style. You make this change in *cboUnique.* That is, you replace *Set tempSymbol = vmr.symbol(i)* with the following code.

1 Replace the previously indicated line with the following.

```
Set tempSymbol = Nothing
   With vmr.symbol(i)
    tempSymbol.SymbolType = vmr.SymbolType
    tempSymbol.Color = .Color
    tempSymbol.Size = .Size
    tempSymbol.Style = .Style
   End With
```

Changing the Symbol in a Flex Grid Cell

Let's look at the issue, raised at the end of the last chapter, of what happens when the user clicks on a symbol in the flex grid. This will fire a click event for the grid. You will know which row and which column have been selected because *grdValues* stores those values in its *row* and *col* properties, respectively. If the column is not the symbol, but the value, the click can be ignored.

If it is the symbol, what should happen? One option would be to provide a pop-up form that would allow the user to change the attributes of a single symbol. This would be very similar to the form for a single-value map. The form should be populated with the values of the *vmr.* How do you know which values to use? Let's take a look at the first part of this code. If you made the change in step 1, your *grdValues_Click* sub should read as follows.

```
Private Sub grdValues_Click()
If tabUp Then
  Exit Sub
End If
If grdValues.Col = 0 Then
  Exit Sub
End If
If grdValues.Row = 0 Then
  Exit Sub
End If
grdValues.Col = 0
curFeatureName = grdValues.Text
grdValues.Col = 1
  Set tempSymbol = Nothing
  With vmr.symbol(grdValues.Row - 1)
    tempSymbol.SymbolType = .SymbolType
    tempSymbol.Color = .Color
    tempSymbol.Style = .Style
    tempSymbol.Size = .Size
  End With
  frmSymbol.Show vbModal
End Sub
```

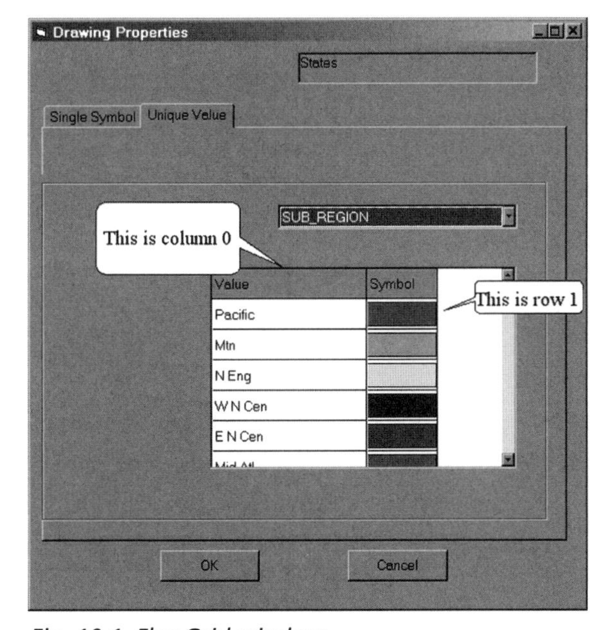

Fig. 10-1. Flex Grid window.

This code first checks to see if *tabUp* is true. If it is true, the sub is exited. It then checks to see if the user clicked in the first column (*grdValues.Col = 0*). If so, the click is ignored (i.e., sub is exited). The program then checks to see if the user clicked in the header row (*grdValues.Row = 0*). Again, if so, the sub is exited. You want the program to react only to clicks on a symbol (see figure 10-1).

If the user clicks on a symbol, the current column is set to column 0. The value in the label field is then copied to *curFeatureName*, and then current column is reset to 1. The code you entered in step 1 sets *tempSymbol* to "nothing," and copies over the values of the *vmr*.

The *grdValues_Click* sub currently ends the *frmSymbol* form being displayed to the user. It is now time to construct that form.

1. Add a new form to your project and name it *frmSymbol*. Give it the controls shown in figure 10-2.

This form is very similar to the single-value map form.

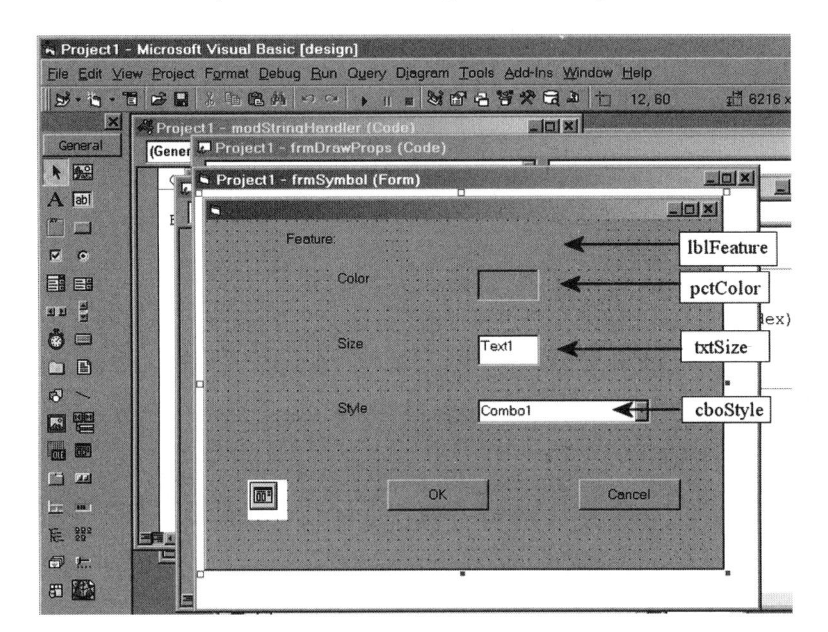

Fig. 10-2. The frmSymbol *form.*

2. In the *frmSymbol*'s code page, add the following *Form_Load* sub.

```
Private Sub Form_Load()
 With drawLayer
   pctColor.BackColor = frmDrawProps.tempSymbol.Color
   lblFeature = frmDrawProps.curFeatureName
   txtSize = frmDrawProps.tempSymbol.Size
   cboStyle.Clear
   Select Case .shapeType
   Case moPoint
      cboStyle.AddItem "Circle"
      cboStyle.AddItem "Square"
      cboStyle.AddItem "Triangle"
      cboStyle.AddItem "Cross"
   Case moLine
      cboStyle.AddItem "Solid Line"
      cboStyle.AddItem "Dash Line"
```

```
      cboStyle.AddItem "Dot Line"
      cboStyle.AddItem "Dash Dot"
      cboStyle.AddItem "Dash Dot Dot"
   Case moPolygon
      Label2.Visible = False
      txtSize.Visible = False
      cboStyle.AddItem "Solid Fill"
      cboStyle.AddItem "Transparent"
      cboStyle.AddItem "Horizontal"
      cboStyle.AddItem "Verical"
      cboStyle.AddItem "Upward Diagonal"
      cboStyle.AddItem "Downward Diagonal"
      cboStyle.AddItem "Cross"
      cboStyle.AddItem "Diagonal Cross"
   End Select
   cboStyle.ListIndex = frmDrawProps.tempSymbol.Style
 End With
End Sub
```

The *frmSymbol* still requires three more subs. The first of these is for when the user clicks on the picture box of a color. When that happens, the program should call a *pctColor_Click* sub. You created such a sub in Chapter 8. All you need to do is copy and paste it into *frmSymbol*'s code page.

3 Copy the *pctColor_Click* sub from *frmDrawProps*' code page and paste it into *frmSymbol*'s code page.

The final subs needed for *frmSymbol* are for its OK and Cancel buttons.

4 Enter the following.

```
Private Sub cmdCancel_Click()
   bolChanged = False
   Unload Me
End Sub
Private Sub cmdOK_Click()
   With frmDrawProps.tempSymbol
      .Color = pctColor.BackColor
      .Style = cboStyle.ListIndex
      .Size = txtSize
   End With
   bolChanged = True
   Unload Me
End Sub
```

Note that both the Cancel and OK subs use a Boolean variable named *bolChanged.* This variable will allow you to check if the user has changed any symbol. That information will be needed in frmDrawProps. Therefore, you will need to give *bolChanged* global scope.

5 In *modUtility,* add the following line.

```
Public bolChanged As Boolean
```

Now that you have created this Boolean variable and managed its value in the OK and Cancel subs, you can use that information in *frmDrawProps.* The *frmDrawProps* code needs to check if the user changed the settings for the symbol (was OK clicked?). If so, the program updates the *vmr,* and then updates the picture in the flex grid.

6 Add the following code to the *frmDrawProps* form's *grdValues_Click* sub, immediately after the line *frmSymbol.Show vbModal.*

```
  If bolChanged = True Then
  With vmr.symbol(grdValues.Row - 1)
   .Color = tempSymbol.Color
   .Style = tempSymbol.Style
   .Size = tempSymbol.Size
  End With
  mapDrawSymbol.TrackingLayer.Refresh True
  mapDrawSymbol.CopyMap 1
  Set grdValues.CellPicture = Clipboard.GetData
  grdValues.CellPictureAlignment = flexAlignCenterCenter
End If
```

Now suppose the user has made these, and possibly other changes, but then decides to discard them by clicking on the Cancel button on the *frmDrawProps* form. What should Cancel on this form (*frmDrawProps*) do? In the previous code, you have changed the *vmr,* but you have not changed the renderer for this layer. If the layer already had a value map renderer (verified by checking the *drawLayer* tag), the program should set the *vmr* back to the layer's renderer. If the layer does not have a renderer, the program simply quits the form (no harm is done). The following *cmdCancel_Click* sub accomplishes these tasks.

7 Enter the following in *frmDrawProps'* code page.

```
Private Sub cmdCancel_Click() 'This is on frmDrawProps code page
  If drawLayer.Tag = "UniqueValue" Then
```

```
    'Need to reset vmr
    Dim i As Integer
    Dim rv As New MapObjects.ValueMapRenderer
    Set rv = drawLayer.Renderer
    vmr.Field = rv.Field
    vmr.SymbolType = rv.SymbolType
    vmr.ValueCount = rv.ValueCount
    For i = 0 To vmr.ValueCount - 1
        vmr.Value(i) = rv.Value(i)
        vmr.symbol(i).Color = rv.symbol(i).Color
        vmr.symbol(i).Style = rv.symbol(i).Style
        vmr.symbol(i).Size = rv.symbol(i).Size
    Next
  End If
  Unload Me
End Sub
```

Your program can now display a value map renderer, manage any editing of its content by the user, and either save or discard those edits, depending on whether the user clicks on the OK or Cancel button when editing the value map renderer.

Initializing frmDrawProps

You now need to figure out how the flex grid will be initialized for cases in which the renderer already exists. First, let's look at what might seem reasonable but does *not* work. Consider the following code. In *Form_Load*, you might try the following.

```
Private Sub Form_Load()
  Set curDrawSymbol.mapControl = Me.mapDrawSymbol
  lblLayerName = drawLayer.Name
  'When first loaded into the map, all layers have a blank tag
  If drawLayer.Tag = "" Then drawLayer.Tag = "SingleSymbol"
  tabUp = True
  Select Case drawLayer.Tag
  Case "SingleSymbol"
      RestoreSingleValueMap
  Case "UniqueValue"
      RestoreUniqueValueMap
  End Select
  Set TabStrip1.SelectedItem = TabStrip1.Tabs.Item(curTab)
  tabUp = False
End Sub
```

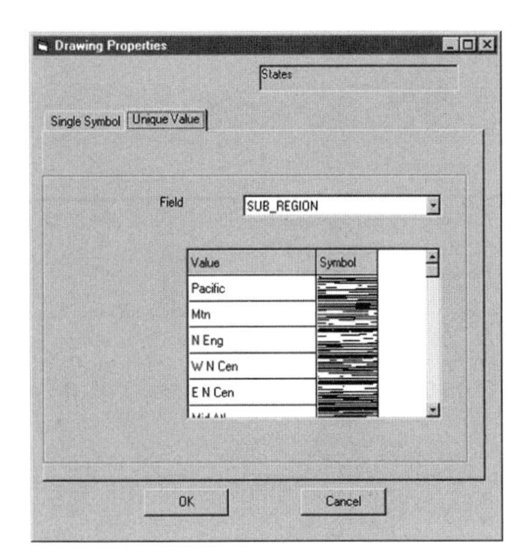

Fig. 10-3. An improperly initialized legend.

This code is designed to load the form and then display the proper tab (and frame) on the screen. That is, if you have a value map renderer for *drawLayer,* you can initialize the form to display it. It seems logical, but your form will look like that shown in figure 10-3 if you use this method.

The pictures in the flex grid are not what they should be. Pictures are placed in the flex grid using the *CopyMap* method to copy the map to the clipboard. You then paste the content of the clipboard into the grid. However, when does this get executed in a case like this? It is during *Form_Load,* which is called *before* the form is displayed on the screen. The result is that the previous code is trying to copy something to the clipboard that is not yet on the screen. To remedy this situation, you need to move the map control from this form (*frmDrawProps*) to the form that calls it (form *Form2,* the layer control form).

Displaying the Current Renderer Page

In order to have *frmDrawProps* display the proper tab and renderer, *mapDrawSymbol* needs to be removed from *frmDrawProps* and placed in form *Form2.*

1. Remove the map control, *mapDrawSymbol,* from the *frmDrawProps* form.
2. Add a map control with the same name (*mapDrawSymbol*) to the layer control form (*Form2*). Hide it behind the list box of layers.
3. Add the following code to form *Form2*'s code page.

```
Private Sub mapDrawSymbol_AfterTrackingLayerDraw(ByVal hDC As
   StdOle.OLE_HANDLE)
  Set frmDrawProps.curDrawSymbol.mapControl = mapDrawSymbol
  Set frmDrawProps.curDrawSymbol.symbol = frmDrawProps.tempSymbol
  frmDrawProps.curDrawSymbol.Draw
End Sub
```

This code sets the class object's (*curDrawSymbol*) map control to the map control on the current form (*Form2*). It also sets its symbol and issues a draw command. The class object *curDrawSymbol* is defined on another form's code page (*frmDrawProps*). Therefore, you have to explicitly reference the form in this code (hence, all the *frmDrawProps* references). Now you need to make some minor changes to the code in *frmDrawProps* code page.

4. Search for every instance of *mapDrawSymbol* and replace it with *Form2.mapDrawSymbol.*
5. Because you no longer have a map control on this form, remove the *AfterTrackingLayerDraw* sub from the *frmDrawProps* code page.
6. Change the *frmDrawProps Form_Load* function to read as follows.

```
Private Sub Form_Load()
  Dim curTab As Integer
  Dim i As Integer
  lblLayerName = drawLayer.Name
  'When first loaded into the map, all layers have a blank tag
  If drawLayer.Tag = "" Then drawLayer.Tag = "SingleSymbol"
  tabUp = True
  Select Case drawLayer.Tag
  Case "SingleSymbol"
      curTab = 1
  Case "UniqueValue"
      curTab = 2
  End Select
  Set TabStrip1.SelectedItem = TabStrip1.Tabs.Item(curTab)
  tabUp = False
End Sub
```

When the form loads, the program checks for the type of renderer. It then sets *TabStrip1*'s *SelectedItem* to that renderer. This fires a *TabStrip1* click event, which makes the proper renderer visible on the form and calls its restore function.

The final steps in implementing this renderer are to change the *RestoreUniqueValueMap* sub so that *frmDrawProps* can be initialized properly if *drawLayer* already has a unique-value map renderer and to make some minor changes to the *cboUnique_Click* sub.

7. Edit the *RestoreUniqueValueMap* sub so that it reads as follows.

```
Private Sub RestoreUniqueValueMap()
  Dim recs As New MapObjects2.Recordset
```

```
   Set recs = drawLayer.Records
   Dim fld As MapObjects2.Field
   Dim i As Integer
   Const tpi = 1440 'Twips per inch
   cboUnique.Clear
   cboUnique.AddItem "None"
   For Each fld In recs.Fields
     If fld.Name <> "Shape" Then
        If fld.Name <> "FeatureId" Then
           cboUnique.AddItem fld.Name
        End If
     End If
   Next
   If drawLayer.Tag <> "UniqueValue" Then
     cboUnique.ListIndex = 0
   Else 'the layer is currently displayed
  'with a vmr
     cboUnique.ListIndex = 0
     Dim rv As New MapObjects2.ValueMapRenderer
     Set rv = drawLayer.Renderer
     vmr.Field = rv.Field
     vmr.SymbolType = rv.SymbolType
     vmr.ValueCount = rv.ValueCount
     For i = 0 To vmr.ValueCount - 1
        vmr.Value(i) = rv.Value(i)
        vmr.symbol(i).Color = rv.symbol(i).Color
        vmr.symbol(i).Style = rv.symbol(i).Style
        vmr.symbol(i).Size = rv.symbol(i).Size
     Next
     For i = 0 To cboUnique.listCount - 1
        If cboUnique.List(i) = rv.Field Then
           cboUnique.ListIndex = i
        End If
     Next
     cboUnique_Click
     Exit Sub
   End If
End Sub
```

The significant changes in this sub from the one in Chapter 9 have to do with a case in which *drawLayer* has a unique-value map renderer. The section of code that corresponds to such a case starts with the *Else* statement. The sub creates a new renderer, *rv*, and sets it equal to the *drawLayer*'s renderer. Next, the properties of that renderer are copied by value to the *vmr* renderer. A *For* loop is then executed to set the selected variable in the *cboUnique* combo box to the *drawLayer* renderer's field. The section of the code ends with a call to *cboUnique_Click*. For *cbo_Unique_Click* to work properly, two small changes need to be made in its code.

8 Remove the following lines from the start of *cboUnique_Click*.

```
If tabUp = True Then
  Exit Sub
End If
```

9 Find the *Do While* loop in *cboUnique sub* and change it to read as follows (the changes are indicated in bold).

```
Do While Not recs.EOF
 uniquevals.Add recs(cboUnique.Text).Value
 If tabUp = True Then
 goOn = True
 Else
 If uniquevals.Count > 25 And goOn = False Then
   retval = MsgBox("There are more than 25 unique _
      values. Continue?", vbYesNo)
   If retval = vbNo Then
   Exit Do
   Else
   goOn = True
   End If
 End If
 End If
recs.MoveNext
Loop
```

The changes to the *cbo_Unique* sub allow the renderer to be initialized properly, even if there are more than 25 unique values.

Summary

The last three chapters have been devoted to renderers, and yet there are many renderers that could be covered. These include classes based on quantiles, standard deviations, and natural breaks, as well as dot density renderers and label renderers. In the next chapter, you will explore one more renderer, the class breaks renderer, based on quantiles.

Chapter 11

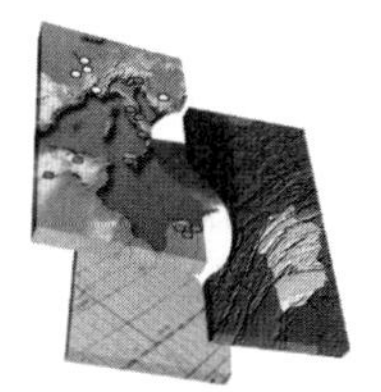

The Quantile Renderer

■■ Introduction

In this chapter you will work with the class breaks renderer. In this type of renderer, you ask the user to select a numeric variable for mapping, and a number of classes. You then assign features to groups based on their variable scores. There are many assignment algorithms you can use, such as equal intervals, quantiles, and standard deviations. Here you will work with quantiles.

■■ Building the Quantile Renderer

CD-ROM NOTE: *The VB project in the* Chapter11_1 *directory on the companion CD-ROM starts here.*

The process for building the quantile renderer is similar to that used for building the unique-value map renderer. You will need to create a tab on the legend editor form for this renderer, create the proper input controls in a frame that will be synchronized with the renderer, and develop a *RestoreQuantileRenderer* sub. You also will need to make modifications to some existing subs in the *frmDrawProps* form. Start by creating a new *Tframe* on the *frmDrawProps* form and a new tab, according to the following parameters.

1. The new tab should have *Quantiles* as its caption and key. The frame should be very similar to the one for the unique-value renderer. It should consist of a combo box for the numeric field names (name it *cboQuantiles* in drop-down list style), a combo box for the number of classes (name it *cboNumclasses* in drop-down list style), and a flex grid for

the quantile classes (name it *grdQuantiles*). Place prompts before each combo box. Finally, add two picture boxes for starting (name it *pctStart*) and ending (name it *pctEnd*) colors of a color ramp (figure 11-1).

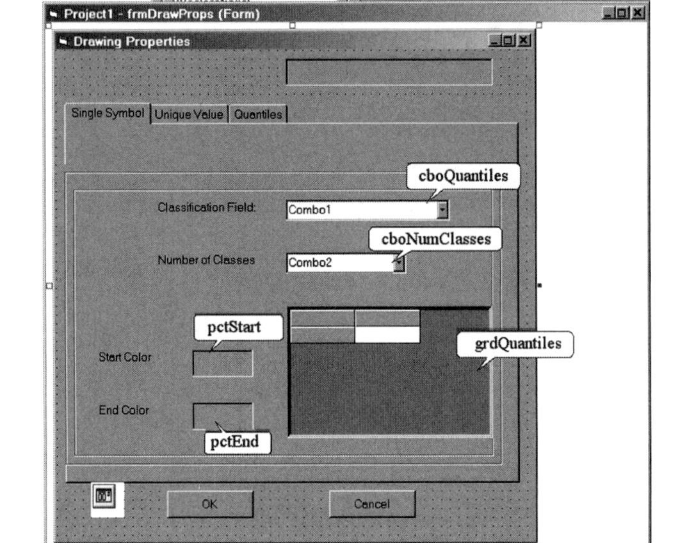

Fig. 11-1. Tframe for Quantiles.

Now you need to make the proper code changes.

2 In *TabStrip1_Click*, add the following case to the *Select Case* section (enter these lines just before the *End Select* statement).

```
Case 3
  tabUp = True
  RestoreQuantileMap
  tabUp = False
```

You now need to write *RestoreQuantileMap*. First consider what you would do if there were no existing quantile renderer. This would be very similar to the bringing up the Unique Value Map frame for the first time. As in that case, you need to set the mapping variable to None and clear the flex grid. For the quantile renderer, you will also need to set default start and end colors and number of classes.

3 Enter the following sub.

```
Private Sub RestoreQuantileMap()
 Dim recs As New MapObjects2.Recordset
 Set recs = drawLayer.Records
```

```
 Dim fld As MapObjects2.Field
 Dim i As Integer
 Const tpi = 1440 'Twips per inch
 cboQuantiles.Clear
 cboQuantiles.AddItem "None"
 For Each fld In recs.Fields
  If fld.Type = moLong or fld.Type = moDouble Then
    If fld.Name <> "FeatureId" then
      cboQuantiles.AddItem fld.Name
    End If
   End If
 Next
 For i = 2 To 10
   cboNumclasses.AddItem i
 Next
 If drawLayer.Tag <> "Quantiles" Then
   cboQuantiles.ListIndex = 0
   cboNumclasses.ListIndex = 3
   'Make 5 the default value
 Else
   'We will develop this below
 End If
End Sub
```

Here, the first *For* loop ensures that the list of variables contains only numeric variables. This is accomplished by the *If* statement that checks that the variable is either a long or a double (see Chapter 3 for the definitions of longs and doubles). The second *For* loop builds a list for the number of groups the user can specify.

At this point, you have populated the two combo boxes, but the flex grid is empty. In the unique-value map renderer, you populated the flex grid when the user chose a variable. Here, you will populate the flex grid when the user chooses a variable or the user changes the number of classes and the variable is not set to *none*. Otherwise, you will leave the grid empty. The following code shows the two functions you need.

4 Enter the following in the *frmDrawProps* code page.

```
Private Sub cboNumclasses_Click()
  If cboQuantiles.ListIndex = 0 Then
    Exit Sub
  End If
  PopulateQuantileGrid
End Sub
Private Sub cboQuantiles_Click()
```

```
    If cboQuantiles.ListIndex > 0 Then
      PopulateQuantileGrid
    End If
  End Sub
```

The first sub is for when the user changes the number of classes; the second sub is for when the user changes the variable. In the first sub, the program checks to see if a variable has been selected. If one has not been selected, it does nothing. The action really takes place in the sub *PopulateQuantileGrid*. To use this, you must first add the following to the top of the *frmDrawProps* code page.

5 Enter the following.

```
Public cbr As New MapObjects2.ClassBreaksRenderer
```

This renderer, *cbr*, is a *ClassBreaksRenderer*. The following are some of its key properties.

- ❒ *BreakCount* is the number of break points. This should be one less than the number of classes (*cboNumclasses – 1*).
- ❒ *Break(j)* is the *j*th break value. This is the upper bound.
- ❒ *Field* is the field on which the breaks are calculated.
- ❒ *SymbolType* is *moPoint*, *moLine*, or *moPolygon*.
- ❒ *Symbol(j)* is the *j*th symbol.
- ❒ *Tag* is the renderer's tag.

You will set *Tag* to *ClassBreaks* when the user commits (clicks on OK) to a quantile renderer. The field value will be the name stored in the *cboQuantiles* combo box. *BreakCount* will be *cboNumclasses – 1*.

The sub *PopulateQuantileGrid* must accomplish several tasks. It has to set the proper number of classes for the renderer and the field on which the drawLayer will be rendered. It then has to assign each feature in the drawLayer to the proper class. It then must populate the grid and set the starting and ending colors of the renderer color ramp. Since this is a long sub, its presentation is broken into steps 6 to 11.

6 Enter the following in the *frmDrawProps* code page.

```
Private Sub PopulateQuantileGrid()
  'Set the number of classes and the variable
  cbr.BreakCount = cboNumclasses.List(cboNumclasses.ListIndex) - 1
  cbr.Field = cboQuantiles.List(cboQuantiles.ListIndex)
```

This sets the number of classes and the classification field for the renderer. These values are taken from the combo boxes on *frm-DrawProps*. You now need to calculate the class break points. If the user wants, for example, four classes, you would try to make the class breaks so that each group has as close to 25% of the cases as possible.

This is not as easy as it seems. There are two potential dangers. First, what do you do with cases that have null values? That is, suppose you were mapping counties, and for the variable in question several counties had null values? Should the 25% be based on all records or only those records that have non-null values? Second, what if the user selects seven classes, but the variable in question has just three values? When there are null records, you need to keep track of the "real" number of cases—cases with non-missing data. When there are more classes than values, you must create only the number of cases for which there are values.

The problem is still a bit more complex than it might appear. Consider a case involving Tennessee. There are 95 counties. If you wanted 10 classes, each group should have approximately 9.5 cases. However, suppose that for the variable in question the first 45 counties all have the same value. What then? In this case, you will have to ignore the 9.5 "cases per group" rule and come as close as possible. The following is the next portion of the *Populate-QuantileGrid* function. Here, you need to declare a few variables and objects.

7 Enter the following.

```
Dim nRecs As Integer
Dim i As Integer
Dim j As Integer
Dim recs As MapObjects2.Recordset
Dim curval As Single
Dim stats As MapObjects2.Statistics
Dim isend As Boolean
Dim nMissing As Integer
Dim nReal As Integer
```

The variables *nMissing* and *nReal* will keep track of the number of missing-value cases and the number of actual cases. That is, *nReal* = *nRecs* – *nMissing*. You now need to get the count of records, select them all, and then calculate the statistics on the field the user has chosen for mapping.

8 Enter the following.

```
nRecs = drawLayer.Records.Count
Set recs = drawLayer.SearchExpression("featureId > -1 order by " & _
   cbr.Field)
recs.MoveFirst
Set stats = recs.CalculateStatistics(cbr.Field)
curval = stats.Min
isend = False
nMissing = 0
```

This code uses the MO *SearchExpression* method to get a sorted list of all records. The search expression has two parts: *featureID > –1*, which will return all records, and *order by*, which will order the records by the *cbr.Field*. The first records will be those with null values. The program then moves to the beginning of the record set with the *recs.MoveFirst* method, and then uses the MO Statistics object to calculate the statistics on this field. The next task is to determine the number of missing cases. The following is the code for determining that number.

9 Enter the following.

```
'Search for the no blank or missing values
  Do While Not recs.EOF
   If recs(cbr.Field).Value = Null Then
    nMissing = nMissing + 1
    recs.MoveNext
   Else
    Exit Do
   End If
 Loop
 nReal = nRecs - nMissing
```

Now you will cycle through all cases. You need to keep track of the current record number (where you are in the records) and whether or not the value of the classification field for the current record differs from the value at which you set the last break.

10 Enter the following.

```
j = 0
For i = 0 To cbr.BreakCount - 1
Do While Not recs.EOF
  j = j + 1
  If j = (i + 1) * Int(nReal / (cbr.BreakCount + 1)) Then
   curval = recs(cbr.Field).Value
  End If
  recs.MoveNext
```

```
  If Not recs.EOF Then
  If j > (i + 1) * Int(nReal / (cbr.BreakCount + 1)) Then
    If recs(cbr.Field).Value <> curval Then
      cbr.Break(i) = recs(cbr.Field).Value
       curval = recs(cbr.Field).Value
     Exit Do
     End If
  End If
  Else
   isend = True
  End If
Loop
If isend = True Then
 cbr.BreakCount = i + 1
 Exit For
End If
Next i
```

This section of code starts with a *For* loop through all break points (*0* to *cbr.BreakCount – 1*). Nested within the *For* loop is a *Do While* loop over all records. Consider the following calculation.

```
(i + 1) * Int(nReal / (cbr.BreakCount + 1))
```

Suppose that *nReal = 100* (100 records) and *BreakCount = 4* (five classes). If you were looking for the first break point, this would evaluate to 1*int(100/5), or 20. Thus, when the twentieth record (j = 20) is reached, the program needs to set *curval* to the value of the classification field for that record.

The program then needs to ascertain if it has moved past the twentieth record. If so, it checks to see if the value of that record is equal to *curval* (the candidate break value). If it is, a case exists in which multiple values lie on the break value.

For example, suppose for the 100-record case the first 30 cases all had the same value. All 30 of them will be assigned to the first quantile. If all 30 cases are not assigned to the first quantile, it is established that the current record must have a value greater than *curval.* The program then needs to set the break value and *curval* equal to the current record's value and break out of the *Do While* loop. The last section of the *For* loop (where *isend = True*) accommodates the situation in which there are more classes than variable values.

At this point, you have calculated the "true" number of class breaks, and you have set the value of each class break point. You now need to populate the flex grid. This code is very much like the code you used to populate the flex grid for the unique-value map renderer.

11 Enter the following.

```
'It is time to build the grid
Const tpi = 1440 'Twips per inch
grdQuantiles.Clear
grdQuantiles.Cols = 2: grdQuantiles.Rows = 2
'Position on first row and set first column settings
grdQuantiles.Row = 0: grdQuantiles.Col = 0
grdQuantiles.ColWidth(0) = tpi * 1.3
grdQuantiles.ColAlignment(0) = 1 'right align
grdQuantiles.Text = "Value"
'Second column settings
grdQuantiles.Col = 1
grdQuantiles.ColWidth(1) = tpi * 0.6
grdQuantiles.Text = "Symbol"
grdQuantiles.FixedRows = 1: grdQuantiles.FixedCols = 0
curFeatureType = drawLayer.shapeType
Select Case drawLayer.shapeType
  Case moPoint
    cbr.SymbolType = moPointSymbol
  Case moLine
    cbr.SymbolType = moLineSymbol
  Case moPolygon
    cbr.SymbolType = moFillSymbol
End Select
cbr.RampColors moPaleYellow, moBlue
'Here is where we set the default color ramp
For i = 0 To cbr.BreakCount
  grdQuantiles.Row = i + 1
  grdQuantiles.Col = 0
  'Now we populate the values column in the flex grid
  If i = 0 Then
    grdQuantiles.Text = stats.Min & " - " & cbr.Break(i)
  Else
    If i >= cbr.BreakCount - 1 Then
      grdQuantiles.Text = cbr.Break(i - 1) & " - " & stats.Max
    Else
      grdQuantiles.Text = cbr.Break(i - 1) & " - " & cbr.Break(i)
    End If
  End If
  'Time to draw the symbol
  grdQuantiles.Col = 1
  Set tempSymbol = Nothing
```

```
    With cbr.symbol(i)
      tempSymbol.SymbolType = cbr.SymbolType
      tempSymbol.Color = .Color
      tempSymbol.Size = .Size
      tempSymbol.Style = .Style
     End With
'The following assumes you put the map control on Form2
'as at the end of chapter 10
     Form2.mapDrawSymbol.TrackingLayer.Refresh True
     Form2.mapDrawSymbol.CopyMap 1
     Set grdQuantiles.CellPicture = Clipboard.GetData
     grdQuantiles.CellPictureAlignment = flexAlignCenterCenter
     grdQuantiles.Rows = grdQuantiles.Rows + 1
     Clipboard.Clear
  Next
  grdQuantiles.Rows = grdQuantiles.Rows - 1
  grdQuantiles.RowHeight(-1) = tpi * 0.25
End Sub
```

This completes the *PopulateQuantileGrid* sub. Now suppose the user chooses the quantile renderer and populates it. What happens when the user clicks on OK in *frmDrawProps*? You need to save the values in the *cbr* to the *drawLayer*'s renderer. The *cmdOK_Click* sub on the *frmDrawProps* code page contains a *Select case* statement, based on the current tab strip selection.

12 Enter the following in *cmdOK_Click* prior to the *End Select* statement.

```
Case 3
  'Check if user clicked OK without settting a variable
  If cboQuantiles.List(cboQuantiles.ListIndex) <> "None" Then
    Dim rc As New MapObjects2.ClassBreaksRenderer
    Set drawLayer.Renderer = rc
    With cbr
      rc.Field = .Field
      rc.SymbolType = .SymbolType
      rc.BreakCount = .BreakCount
      rc.RampColors moPaleYellow, moBlue
      For i = 0 To .BreakCount - 1
        rc.Break(i) = .Break(i)
        rc.symbol(i).Style = .symbol(i).Style
        rc.symbol(i).Size = .symbol(i).Size
      Next
      drawLayer.Tag = "Quantiles"
    End With
  End If
```

Restoring the Quantile Renderer

CD-ROM NOTE: *The VB project in the* Chapter11_2 *directory on the companion CD-ROM starts here.*

If you run this program as is, it will work—sort of. You can map a layer with a quantile map. However, if you go back to edit the quantile map, that will not work. If you look closely at *RestoreQuantileMap*, you will see a comment you never completed in the *Else* portion of *If drawLayer.Tag <> "Quantiles"*. It is time to supply the code for implementing the color ramp pictures.

1 Perform the following in Restore Quantile Map.

Start by changing

```
Else
' We will develop this below
End If
```

to

```
Else
 Dim rc As New MapObjects2.ClassBreaksRenderer
 Set rc = drawLayer.Renderer
 cbr.Field = rc.Field
 cbr.SymbolType = rc.SymbolType
 cbr.BreakCount = rc.BreakCount
 For i = 0 To rc.BreakCount
   cbr.symbol(i).Style = rc.symbol(i).Style
   cbr.symbol(i).Size = rc.symbol(i).Size
 Next
 pctStart.BackColor = rc.symbol(0).Color
 pctEnd.BackColor = rc.symbol(rc.BreakCount).Color
 For i = 0 To cboQuantiles.listCount - 1
   If cboQuantiles.List(i) = rc.Field Then
     cboQuantiles.ListIndex = i
   End If
 Next
 For i = 0 To cboNumClasses.listCount - 1
   If cboNumClasses.List(i) = rc.BreakCount + 1 Then
     cboNumClasses.ListIndex = i
   End If
 Next
End If
```

This will initialize the *cbr* renderer to that of the *drawLayer*. It also will set the proper values in the starting and ending color picture boxes, and set the *ListIndex* for the number of classes and classification field combo boxes. However, you need to be careful. Every time you set a *ListIndex*, you fire a click event.

2 Change the click event for *cboQuantiles* to the following (the change to the existing code is indicated in bold).

```
Private Sub cboQuantiles_Click()
 If cboQuantiles.ListIndex > 0 Then
  If cboNumClasses.ListIndex >= 0 Then 'This if is new
   PopulateQuantileGrid
  End If
 End If
End Sub
```

The final step is to work with the starting and ending colors for the color ramp.

3 In *RestoreQuantileMap*, make the following changes (indicated in bold).

```
If drawLayer.Tag <> "Quantiles" Then
 cboQuantiles.ListIndex = 0
 cboNumClasses.ListIndex = 3
 'Make 5 the default value
 pctStart.BackColor = moPaleYellow
 pctEnd.BackColor = moBlue
Else
```

4 In *cmdOK_click*, change how the color ramp is set for Case 3 to the following.

```
rc.RampColors pctStart.BackColor, pctEnd.BackColor
```

Finally, you want to be able to change the starting and ending colors. Therefore, you will add click events for the two picture boxes.

5 Enter the following.

```
Private Sub pctEnd_Click()
 Dim curColor As Long
 CommonDialog1.CancelError = True
 On Error GoTo ErrHandler
 CommonDialog1.ShowColor
 curColor = CommonDialog1.Color
 pctEnd.BackColor = curColor
 PopulateQuantileGrid
 Exit Sub
```

```
ErrHandler:
'Do nothing,just exit
 Exit Sub
End Sub
Private Sub pctStart_Click()
 Dim curColor As Long
 CommonDialog1.CancelError = True
 On Error GoTo ErrHandler
 CommonDialog1.ShowColor
 curColor = CommonDialog1.Color
 pctStart.BackColor = curColor
 PopulateQuantileGrid
 Exit Sub
ErrHandler:
'Do nothing,just exit
 Exit Sub
End Sub
```

Summary

The last four chapters have dealt with renderers. In these chapters you learned how to create and manipulate a Tab Strip control, how to work with flex grids, and how to create renderers for single-value maps, unique-value maps, and quantile maps. However, there are many other options you could have incorporated into the coding scenarios you have been working with. The following are options you might want to explore.

- ❐ Dot density renderers
- ❐ Statistical classification, such as on standard deviations
- ❐ Equal interval classification
- ❐ Natural break classification
- ❐ Text renderers

We could spend several more chapters on rendering. However, between the material in these chapters and the code in *MoView2*, you should be able to figure out the remaining cases.

Chapter 12

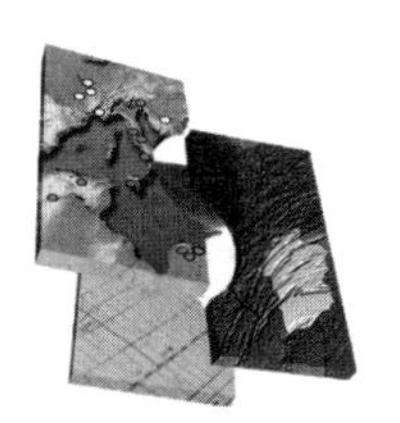

Collections, Classes, and Advanced Selections

Introduction

If you think back to when you implemented the Identify button (Chapter 7), you will recall that you performed the following.

- Created a MapObjects record set.
- Used *TrackRentangle* to track mouse movements, and used the *SearchShape* method to see if any features were within the rectangle.
- If no features were within a rectangle (particularly if the user simply clicked on the map), you used *SearchbyDistance* to see if the user clicked on or near a feature in the current layer.
- Drew the selected features in a special color and displayed their attributes.

Recall also that every time the Identify button was used, a new *Recordset* object was generated. As you have probably guessed, there is more you can do with selected sets. Sometimes you want to add to the currently selected set. Sometimes you want to select from the selected set. You may want to save the selected features to a new shape. Finally, you may want to select features in one layer based on their relationship with features in another layer (overlay methods). In this chapter, you will implement the "add-to-set" option, the "save selection" option, and a form of polygon overlay.

The Collection Object and the Selection Button

Collections are special VB structures that help you maintain lists of objects. You have been dealing with collections since Chapter 4, in which you created a map layer collection. The following methods are associated with collections.

- *Add:* Adds an object to the collection.
- *Count:* Returns the number of items in the collection.
- *Item:* Returns an item. This is indexed by number and, optionally, a key value.
- *Remove:* Removes an item from a collection.

Collections are much like arrays, but are easier than arrays to maintain. In VB, you do not need to resize the collection every time you add or remove an item. VB takes care of that overhead for you.

The strategy you will use is to create a collection of selected records. Suppose, for example, the user selects *n* records. In such a case, you could place all selected records into the collection as a single item (a single *Recordset* element in the collection consisting of *n* records). In essence, here you would be creating a collection, each element of which is also a collection. Alternatively, you could add each record as a single item in a collection. Thus, you would add *n* items to the collection. Either method will serve the purpose. Here, you will use the latter strategy.

CD-ROM NOTE: *The VB project in the* Chapter12_1 *directory on the companion CD-ROM starts here.*

1. In form *Form1*'s code page, add the following statement at the top.

```
Public colRecSet As New Collection
```

This collection will hold all selected records. Next, you need to add an image to the image *ImageList* control.

2. Right-click on the ImageList control to bring up its properties dialog. Add the image *spatial select.bmp.* (This is found in the *MoView2 Bitmaps* directory.)

3 Once the image is added to the list, add a button to form *Form1*'s toolbar that uses this image. Place this button just to the right of the Identify button (Index = 7).

4 Name the added button Select and make it part of the button group (see figure 12-1).

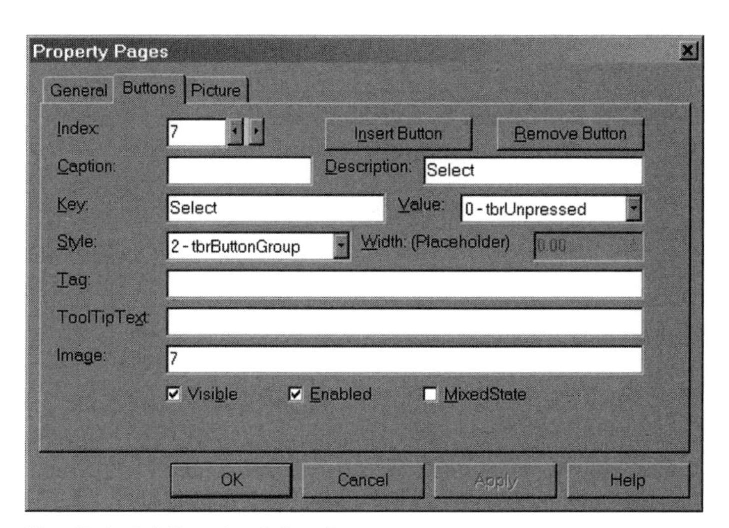

Fig. 12-1. Adding the Select button.

You now have a button for spatial selection. When should this button be available for use? That is, when can the user select records? If you think about it, the Selection tool should be available whenever the Identify tool is available. Further, whenever the Identify tool is turned off, the Selection tool should be turned off.

5 Search form *Form1*'s code page, and wherever the Identify button is enabled, set the Selection tool's button enable property to True. Similarly, wherever the Identify button is disabled, set the Selection tool's enable property to False.

Enabling Selections

Now that you have a selection button, you need to figure out how it should behave. If the user selects features with the mouse and no key is pressed, the program should create a new selected set. If the Shift key is pressed, the selected set should be added to. If any other key is pressed, the operation should be ignored.

If you look at the *MouseDown* event, you will see that one of the arguments passed in is *Shift As Integer.* When no key is pressed, then *Shift = 0.* When the Shift key is pressed, *Shift = 1.* You will use these values to distinguish between the two cases.

1 At the start of the *Map1_MouseDown* sub, add the following.

```
Dim i as Integer
```

2 Add the following code immediately after the last *End If* in the Identify case.

```
ElseIf Toolbar1.Buttons("Select").Value = 1 Then
  Set r = Map1.TrackRectangle
  Set recs = ActiveLayer.SearchShape(r, moAreaIntersect, "")
  If recs.EOF Then
    tol = Map1.ToMapDistance(100)
    Set pt = Map1.ToMapPoint(X, Y)
    Set recs = ActiveLayer.SearchByDistance(pt, tol, "")
  End If
  If recs.Count > 0 Then
    Set gSelection = recs
  Else
    Set gSelection = ActiveLayer.SearchExpression("featureId = -1")
  End If
  gSelection.MoveFirst
```

This code begins by selecting the proper case (*Toolbar.Buttons("Select").Value = 1*). The remaining lines of the code use the same approach as in the Identify button to get the selected features.

Now that the selected records are assigned to *gSelection*, you need to determine if the user wants to create a new set or add to an existing selected set. Remember that the collection *colRecSet* is being used to contain selected records. If there are any items in *colRecSet* and *Shift* = 0, the records in *colRecSet* must be eliminated before the records in *gSelection* can be added to the collection.

3 Continuing in the *MouseDown* event sub, enter the following immediately after the code you entered in step 2.

```
If (Shift = 0) Then
   For i = 1 To colRecSet.Count
     colRecSet.Remove (1)
   Next i
   Dim aRec As MapObjects2.Recordset
   Do While Not gSelection.EOF
     Set aRec = ActiveLayer.SearchExpression("featureId = " & _
         gSelection.Fields("FeatureID").ValueAsString)
     colRecSet.Add aRec
     gSelection.MoveNext
   Loop
End If
```

This code starts by checking the status of *Shift*. If *Shift* is zero (no key is pressed), the program removes all elements in the *colRecSet* collection. Note that the first element is always removed. This is because every time you remove a record, the index of the last record changes. However, the first record's index never changes. By removing only the first record, you will never specify a record index that is out of range. The program then searches the *gSelection* record set, selecting each record one at a time. This is done by selecting each record on its feature ID. Each record is saved to a record set variable called *aRec*, which is added to the collection. The program then moves on to the next record.

If the user holds the Shift key down, the program needs to add the elements of *gSelection* to the currently selected set. The program cannot simply add all records in *gSelection* to the collection, because a newly selected record might already be in the selected set. In such cases, you do not want the program to double count. The following code provides for the situation in which *Shift = 1*.

4 Enter the following immediately after the code you entered in step 3.

```
If (Shift = 1) Then 'Add to set
  Dim found As Boolean
  Do While Not gSelection.EOF
    found = False
    Dim bRec As MapObjects2.Recordset
    For i = 1 To colRecSet.Count
      Set bRec = colRecSet.Item(i)
      If (bRec.Fields("FeatureID").Value = gSelection.Fields _
         ("FeatureID").Value) Then
      found = True
      Exit For
     End If
    Next i
    If (Not found) Then 'add new record
     Set aRec = ActiveLayer.SearchExpression("featureId = " & _
         gSelection.Fields("FeatureID").ValueAsString)
     colRecSet.Add aRec
    End If
    gSelection.MoveNext
  Loop
End If
Map1.Refresh
```

In this code, when *Shift = 1*, a Boolean called *found* is created to see if any of the newly selected features are already in the selected set. The program then proceeds through a double loop. First, it grabs a record from the features the user just selected. It then goes through the collection to see if that record is already in the set. If it is, *found* is set to True, and the *For* loop over the collection is exited. This is done because once a record is found in the collection, there is no need to continue to look for it. Once the program gets beyond the inner *For* loop, it checks *found*. If *found* is True, the program simply gets the next member of the *gSeletion* record set. If found is *not* True, the record is added to the collection.

The last step is to refresh the map. Refreshing the map calls the *AfterLayerDraw* sub. Think for a moment about this. Suppose the user selects a set and then zooms in on an area. Do you want the selected set to be highlighted? As you currently have the *AfterLayerDraw* sub written for the Identify button, any redraw where the button pressed is not Identify wipes out the highlighting of features. This may be acceptable for Identify, but it is probably not what you want for selected sets. The following code corrects the situation by changing the *AfterLayerDraw* function.

5 Edit *AfterLayerDraw* to read as follows.

```
Private Sub Map1_AfterLayerDraw(ByVal index As Integer, ByVal _
    canceled As Boolean, ByVal hDC As StdOle.OLE_HANDLE)
 Dim sym As New MapObjects2.symbol
 Dim i As Integer
 Dim bval As MapObjects2.Recordset
 sym.Color = moYellow
 For i = 1 To colRecSet.Count
   Set bval = colRecSet.Item(i)
   Map1.DrawShape bval("Shape").Value, sym
 Next i
 If (Toolbar1.Buttons("Identify").Value = tbrPressed) Then
  If gSelection Is Nothing Then
   Exit Sub
  End If
  If index > 0 Then
   Exit Sub
  End If
  sym.Color = moRed
  gSelection.MoveFirst
  Do While Not gSelection.EOF
   Map1.DrawShape gSelection("Shape").Value, sym
   gSelection.MoveNext
  Loop
 End If
End Sub
```

This code has added a new record set object, *bval*, to the declarations. The selected symbol color is set to *moYellow* (you could use a different color). Every element in the selected set is drawn in yellow. If there are no elements, nothing gets drawn. Note that this works no matter what button in the button bar is pressed. That is, if the user zooms in on an area, the selected features will still be drawn in yellow.

This sub ends by checking to see if the pressed button in the toolbar is the Identify button. If it is, the program proceeds as before by highlighting the identified records last. Exercise 12-1, which follows, provides you with an opportunity to practice specifying selection from the selected set.

Adjunct Exercise 12-1: Specifying Selection from the Selected Set

If the user presses the Alt key, the value of *Shift* is 2. Implement a procedure to select from the selected set when this occurs.

The Need for a Class

***CD-ROM NOTE:** The VB project in the* Chapter12_2 *directory on the companion CD-ROM starts here.*

If you try the program as it now exists (having one map layer), it will work. However, if there is more than one layer in the map, there could be problems. Suppose you had two layers, *cities* and *states*. You make *states* the current layer, and select a few states. Everything is fine. Next, you make *cities* the current layer, and select some cities. What happens? The cities will be selected, but the *states* selection will be wiped out. Suppose you held the Shift key down? Well, the states and cities will be selected, but the collection will be a mix of selected states and selected cities. How would you save the selected features of just one of these?

To handle this type of situation, you need to create a class. Recall from Chapter 1 that classes encapsulate properties and methods. For this selection problem, you want to keep track of selected records by layer. It would make sense, therefore, to have a layer name property and a collection property (to keep track of the selected records, as you did previously). You need to be able to access the layer name and the record set collection. The following steps show how to build the class.

1 From the VB menu, Select Project > Add Class Module.
2 Select a standard Class Module, and name the new module *clsRecSet.*
3 Enter the following in the new module's code page.

```
Option Explicit
Private strLName As String
Dim lyrRecSet As New Collection
Public Property Get Name() As String
 Name = strLName
End Property
Public Property Let Name(curlname As String)
 strLName = curlname
End Property
Public Property Get RecSetCol() As Collection
 Set RecSetCol = lyrRecSet
End Property
```

Each instance of this class will contain a string, called *strLName* (for string layer name), and a collection called *lyrRecSet,* which will contain the named layer's selected records collection. The *Get* method for the name returns the value of *strLName.* The record set, *RecSetCol,* is set equal to the class's collection. The *Let* function places the name in *lyrLName.* Notice that *lyrRecSet* is defined as *New.* This obviates the need for a *Set* property method for *LyrRecSet.* All other methods you need (*Add, Remove, Item,* and *Count*) are already parts of collections, so there is no need to redefine them here.

▪▪ Using the Class

You need to form a collection, as you did previously. However, this collection will be a collection of members of the class *clsRecSet.* Each member of this class will have a layer name and collection. Whenever the user chooses from a layer for the first time, the program will create a new instance of the collection. (Alternatively, you could design the program to create a new instance of the class every time a layer is added to the map.) To achieve this, you need to add selected records to the proper element in the class (i.e., to the correct layer) collection (*lyrRecSet*).

1 At the top of form *Form1*'s code page, replace

```
Public colRecSet As New Collection
```

with

```
Public colRecSetClass As New Collection
```

In the *MouseDown* event, the program needs to check to see if the current button is the selection button. If it is, the program then needs to either create a new element in the *colRecSetClass* collection or update the selected records in an existing element in that collection.

2 Edit the *MouseDown* event sub to read as follows.

```
ElseIf Toolbar1.Buttons("Select").Value = 1 Then
  curIndex = -1
  Dim bLayer As New clsRecSet
  If (colRecSetClass.Count = 0) Then
   bLayer.Name = ActiveLayer.Name
   colRecSetClass.Add bLayer
   curIndex = 1
  Else
   For i = 1 To colRecSetClass.Count
    If (colRecSetClass.Item(i).Name = ActiveLayer.Name) Then
     curIndex = i
     Exit For
    End If
   Next i
  End If
  If (curIndex = -1) Then
   bLayer.Name = ActiveLayer.Name
   colRecSetClass.Add bLayer
   curIndex = colRecSetClass.Count
  End If
```

This code sets an integer, *curindex*, to –1. This will store the index number of the instance of *clsRecSet* that needs to be manipulated. This code also creates a new instance (named *bLayer*) of the class *clsRecSet*. If there are no instances of *clsRecSet* in the collection of layers with selections (*colRecSetClass.Count = 0*), a new instance of *clsRecSet* is added the collection *colRecSetClass*. This is done by setting *bLayer*'s name to the active layer's name, and then adding it (*bLayer*) to the collection.

If there are members of *colRecSetClass*, the program searches to see if the active layer has been previously added to the collection. If it has, the program gets its index number. If, after searching all members in the *colRecSetClass* collection the program cannot find *ActiveLayer's* name (*curIndex = -1*), the name of *bLayer* is set to that of the active layer and is added to the collection *colRecSetClass*. At this point, the coding is similar to previous code, except that it needs to get the current class's record set. The following code achieves this.

3 In the *MouseDown* event sub, locate the line *gSeletction.MoveFirst* in the *Select* button case. Enter the following after that line.

```
If (Shift = 0) Then
  For i = 1 To colRecSetClass.Item(curIndex).RecSetCol.Count
    colRecSetClass.Item(curIndex).RecSetCol.Remove (1)
  Next i
  Do While Not gSelection.EOF
    Dim aRec As MapObjects2.Recordset
    Dim curvalue As String
    Set aRec = ActiveLayer.SearchExpression("featureId = " & _
        gSelection.Fields("FeatureID").ValueAsString)
    colRecSetClass.Item(curIndex).RecSetCol.Add aRec
    gSelection.MoveNext
  Loop
End If
If (Shift = 1) Then 'Add to set
  Do While Not gSelection.EOF
    Dim found As Boolean
    found = False
    For i = 1 To colRecSetClass.Item(curIndex).RecSetCol.Count
     Dim bRec As MapObjects2.Recordset
     Set bRec = colRecSetClass.Item(curIndex).RecSetCol.Item(i)
     If (bRec.Fields("FeatureID").Value = gSelection.Fields _
         ("FeatureID").Value) Then
       found = True
       Exit For
     End If
  Next i
  If (Not found) Then 'new record
     Set aRec = ActiveLayer.SearchExpression("featureId = " & _
         gSelection.Fields("FeatureID").ValueAsString)
     colRecSetClass.Item(curIndex).RecSetCol.Add aRec
    End If
    gSelection.MoveNext
  Loop
End If
```

Note in this code how the current element's collection (*colRecSetClass.Item(curIndex).RecSetCol*) is accessed, and how each record in that collection (*colRecSetClass.Item(curIndex).RecSetCol.Item(i)*) is accessed. *colRecSetClass.Item(curIndex)* returns an element in the collection *colRecSetClass.* That element is an instance of the class *clsRecSet.* The *RecSetCol* property of that class returns the collection of selected records for that element, and *RecSetCol.Item(i)* returns the *i*th selected record in that collection of selected records. In order to draw each layer's selected features, if there are any, you will need to make the following changes to *AfterLayerDraw.*

4 Edit *AfterLayerDraw* so that the beginning of that sub reads as follows.

```
Private Sub Map1_AfterLayerDraw(ByVal index As Integer, ByVal _
    canceled As Boolean, ByVal hDC As StdOle.OLE_HANDLE)
  Dim sym As New MapObjects2.symbol
  Dim i As Integer
  Dim bval As MapObjects2.Recordset
  Dim sellayer As Integer
  sellayer = -1
  For i = 1 To colRecSetClass.Count
   If (Map1.Layers(index).Name = colRecSetClass.Item(i).Name) Then
    sellayer = i
    Exit For
   End If
  Next i
  If (sellayer >= 0) Then
   sym.Color = moYellow
   For i = 1 To colRecSetClass.Item(sellayer).RecSetCol.Count
    Set bval = colRecSetClass.Item(sellayer).RecSetCol.Item(i)
    Map1.DrawShape bval("Shape").Value, sym
  Next i
 End If
```

After each layer is drawn, *AfterLayerDraw* checks to see if that layer has any selected features. If it does, *sellayer* is set to the index number of the most recently drawn layer's position in the *colRecSetClass* collection. Each selected element is then drawn with the selection color. The rest of this sub (for the Identify button) is unchanged. Now you can select from many layers, keeping each layer's selected set in its own class.

Saving the Selected Set

You will often make a selection with the goal of creating a new shape file of the selected elements. In many ways, this is like adding a shape file to a map control, only in reverse. Some of the same tools you used in adding a shape file are used in saving the active layer's selected features.

However, what if the active layer has no se-lected features? In this case, you save the entire layer.

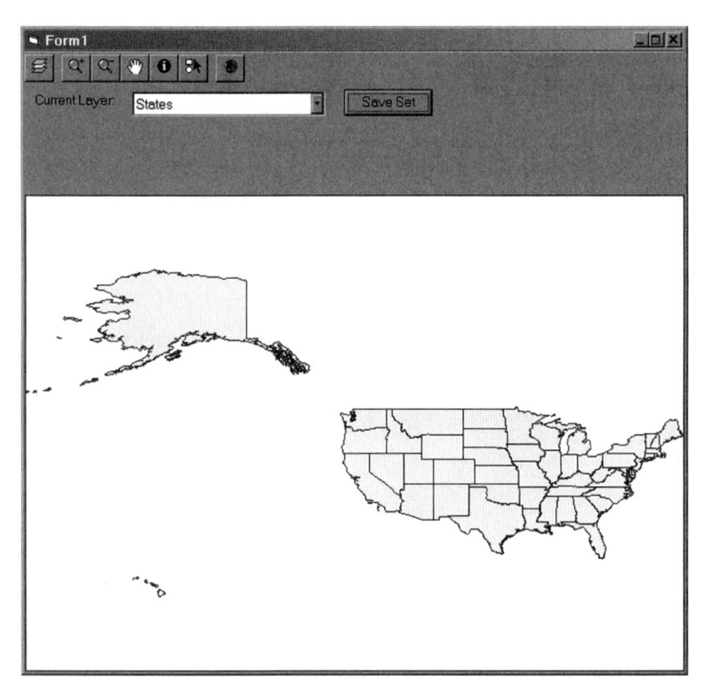

Fig. 12-2. Save Set button.

1. Add a command button to form *Form1* to the right of the combo box for selecting the active layer.
2. Name this button *cmdSave*, and give it the caption *Save Set* (see figure 12-2).

This button should be enabled only when there is an active layer. This is precisely when the Identify and Select buttons are enabled. Therefore, wherever the following statements exist, the statement *cmdSave.Enabled = True* needs to be added.

```
Toolbar1.Buttons("Identify").Enabled = True
Toolbar1.Buttons("Select").Enabled = True
```

3. Add the *cmdSave.Enabled = True* statement where necessary.
4. Disable the *cmdSave* button wherever the Identify and Select toolbar buttons' *Enable* values are set to False.

Now you can start building the *cmdSave_Click* event's code.

5. Enter the following in form *Form1*'s code page.

```
Private Sub cmdSave_Click()
  Dim fullFile As String, path As String, tempChar As String, _
      ext As String
  Dim Test As Boolean
  Dim textPos As Long, periodPos As Long
  Dim curPath As String
  'Execute common dialog for setting a file to save.
  Dim strShape As String
  strShape = "Shape files (*.shp)|*.shp"
  CommonDialog1.Filter = strShape
  CommonDialog1.Flags = cdlOFNOverwritePrompt
  CommonDialog1.CancelError = True
  On Error GoTo ErrHandler
  CommonDialog1.DialogTitle = "Set file for saving layer"
  CommonDialog1.ShowSave
  If CommonDialog1.filename = "" Then Exit Sub
```

This code begins by dimensioning some required strings. It then sets the *Filter* (shape files), *Flags* (prompt if shape is going to overwrite), *CancelError*, and *DialogTitle* parameters for a Save As common dialog. If the user selects nothing, the program exits the sub. If the user clicks on the Cancel button, the program goes to the *ErrHandler*, which will also exit the sub.

The common dialog will return a long string, which includes the path and file name the user has set. The following section of code parses this into path and file name.

6 Continuing with the *cmdSave_Click* event, enter the following.

```
fullFile = Trim$(CommonDialog1.filename)
textPos = Len(fullFile)
Test = False
'This loop goes backward through the string, searching for the
'last back slash. This marks the base path from the returned string.
Do While Test = False
  textPos = textPos - 1
  tempChar = Mid$(fullFile, textPos, 1)
  If tempChar = "." Then
    periodPos = textPos
  ElseIf tempChar = "\" Or textPos = 0 Then
    Test = True
  End If
Loop
'Path is the part of the full file string up to the last back slash.
curPath = Left$(fullFile, textPos - 1)
```

```
'Send the file name to the procedures that save the layers...
Dim filename As String
filename = Left(CommonDialog1.FileTitle, Len(CommonDialog1.FileTitle) - 4)
```

Once you know the path and the file name, you can set the *dataconnection* object (see Chapter 5 for information about *dataconnection* objects and *geodataset* objects).

7 Enter the following directly below the code entered in step 6.

```
Dim dCon As New DataConnection
Dim gSet As GeoDataset
dCon.Database = curPath
If Not dCon.Connect Then
  MsgBox "no connection"
  Exit Sub
End If
Dim tabDesc As New MapObjects2.TableDesc
Dim fldsList As MapObjects2.Fields
Dim geoDS As New MapObjects2.GeoDataset
Dim recs1 As New MapObjects2.Recordset
Dim allrecs As New MapObjects2.Recordset
Dim newlyr As New MapObjects2.MapLayer
Dim sellayer as Integer
Dim I as Integer
Dim fld as MapObjects2.Field
Set allrecs = ActiveLayer.Records
Set tabDesc = ActiveLayer.Records.TableDesc
```

This code creates a new *dataconnection* object and geodata set. It then dimensions some necessary new objects, and sets the record set *allrecs* equal to the entire active layer's record set. It also sets *tabDesc*, a table description object, to the active layer's *TableDesc*.

You may recall that you were asked in Chapter 7 to find the number of fields in a *recordset* object and in a *tabledesc* object. These values differed by two. That is, the MapObjects *recordset* object had two more fields than the *TableDesc* object. These fields are the *FeatureID* and *Shape* fields. These fields are ESRI's way of connecting a *dbf* file to the geographic coordinates in a shape file. To create a standard *dbf* file, you need a data dictionary of the fields you want to add to that file. That data dictionary is what is in the *TableDesc* object. Once you have the *TableDesc*, you can create a geodata set for the correct type of geographic feature (point, line, or polygon).

8 Continuing with the *cmdSave_Click* sub, enter the following.

```
Select Case ActiveLayer.shapeType
Case 21:
  Set geoDS = dCon.AddGeoDataset(filename, moPoint, tabDesc)
Case 22:
  Set geoDS = dCon.AddGeoDataset(filename, moLine, tabDesc)
Case 23:
  Set geoDS = dCon.AddGeoDataset(filename, moPolygon, tabDesc)
End Select
Set newlyr.GeoDataset = geoDS
Set recs1 = newlyr.Records
```

You have now determined the proper shape type and have created an empty shape with the proper table structure (*TableDesc*) for that shape type (point, line, or polygon). The program needs to see if there are any selected features for the active layer. If there are, the selected features are saved. If not, the entire layer is saved.

9 Enter the following directly below the code entered in step 8.

```
sellayer = -1
For i = 1 To colRecSetClass.Count
If (ActiveLayer.Name = colRecSetClass.Item(i).Name) Then
  sellayer = i
  Exit For
End If
Next i
If (sellayer = -1) Then
  'Dump the entire layer
  allrecs.MoveFirst
  Do While Not allrecs.EOF
    recs1.AddNew
    For Each fld In allrecs.Fields
      recs1.Fields(fld.Name).Value = fld.Value
    Next fld
    recs1.Update
    allrecs.MoveNext
  Loop
  Else 'save selected
  For i = 1 To colRecSetClass.Item(sellayer).RecSetCol.Count
    recs1.AddNew
   For Each fld In colRecSetClass.Item(sellayer).RecSetCol.Item(i).Fields
     recs1.Fields(fld.Name).Value = fld.Value
    Next fld
    recs1.Update
  Next i
End If
```

If you want, you can automatically add the new shape to the map. Alternatively (and better), create a Yes/No dialog asking the user if he/she wishes to add the new shape to the map.

There is one last issue unresolved. Suppose the user selects some features from a layer, such as *states.shp*. The user then attempts to save these to the same name in the same directory. As the program is written now, this will cause an error, and the original shape file will be corrupted. The following represents one way of avoiding this.

10 In form *Form2*'s *addShape* sub, find the following line.

```
Form1.Map1.Layers.Add newLayer 'Add MapLayer to Layers collection
```

11 Immediately after the line entered in step 10, add the following.

```
newLayer.Tag = basepath & "\" & shpfile & ".shp" & "|" & ""
```

Because the program allows the user to add shapes only, you know that the extension will be *.shp*. You use this knowledge, along with the path and file name, to set the layer's tag. However, you have to be careful. The *frmDrawProps* also uses the tag to store information. You need to concatenate the file name (with full path) and the renderer name. Whenever a shape is added, its renderer name will be set to *""*. The vertical bar (|) is used as a token to separate the file name from the renderer name. The file name lies to the left of the vertical bar; the renderer name lies to its right.

12 In form *Form1*'s *cmdSave_Click* sub, make the following changes to prohibit the overwriting of an open shape file. Locate the line that sets the value of fullFile.

```
fullFile = Trim$(CommonDialog1.filename)
```

13 Immediately after the line added in step 12, add the following.

```
Test = False
textPos = Len(ActiveLayer.Tag) + 1
Do While Test = False
  textPos = textPos - 1
  tempChar = Mid$(ActiveLayer.Tag, textPos, 1)
  If tempChar = "|" Then
    Test = True
  End If
Loop
If UCase(fullFile) = UCase(Left$(ActiveLayer.Tag, textPos - 1)) Then
  MsgBox "error-cannot save to the name of an open shape"
```

```
    Exit Sub
End If
```

The rest of this sub is unchanged. The previous code searches the active layer's tag for the path and file name for that layer's shape file. It then compares that value to the path and file name to which the user is trying to write the saved the selected records. If these are the same, an error message is displayed and the program exits the sub.

Because the strategy implemented previously changed the structure of tag property for each layer, you need to make some changes to the *frmDrawProps* code to accommodate the new tag structure.

14 At the top of *frmDrawProps* code page, add the following declaration.

```
Public renderPos As Long
Public curRenderName As String
```

In the following, you will use the *renderPos* value to store the location of the vertical bar, and the *curRendername* string to store the name of the current renderer. The following code gets these values every time the form is loaded.

15 Edit *frmDrawProp*'s *Load* sub to read as follows. Changes to *Form_Load* are indicated in bold

```
Private Sub Form_Load()
 Dim curTab As Integer
 Dim i As Integer
 lblLayerName = drawLayer.Name
 Dim tempChar As String
 Dim Test As Boolean
 Test = False
 renderPos = Len(drawLayer.Tag) + 1
 Do While Test = False
  renderPos = renderPos - 1
  tempChar = Mid$(drawLayer.Tag, renderPos, 1)
  If tempChar = "|" Then
   Test = True
  End If
 Loop
 curRenderName = Right$(drawLayer.Tag, Len(drawLayer.Tag) - renderPos)
 tabUp = True
 If curRenderName = "" Then
  curRenderName = "SingleSymbol"
 End If
```

```
 Select Case curRenderName
 Case "SingleSymbol"
  curTab = 1
 Case "UniqueValue"
  curTab = 2
 Case "Quantiles"
  curTab = 3
 End Select
 Set TabStrip1.SelectedItem = TabStrip1.Tabs.Item(curTab)
 tabUp = False
End Sub
```

You need to change the way new renderer information is saved when the user clicks on the OK button.

16 Edit the *cmdOK_Click* sub so that it reads as follows. Changes to the previous version of this sub are indicated in bold.

```
Private Sub cmdOK_Click()
 Dim pathName As String
 Select Case TabStrip1.SelectedItem.index
 Case 1
  Set drawLayer.Renderer = Nothing
  curRenderName = ""
  pathName = Left$(drawLayer.Tag, renderPos)
  drawLayer.Tag = pathName + curRenderName
  With drawLayer.symbol
   .Color = pctColor.BackColor
   .Style = cboStyle.ListIndex
   If drawLayer.shapeType <> moPolygon Then
    .Size = txtSize
   End If
  End With
 Case 2
  'Check if user clicked OK without settting a variable
  If cboUnique.List(cboUnique.ListIndex) <> "None" Then
   Dim rv As New MapObjects2.ValueMapRenderer
   Set drawLayer.Renderer = rv
   rv.Field = vmr.Field
   rv.SymbolType = vmr.SymbolType
   rv.ValueCount = vmr.ValueCount
   Dim i As Integer
   For i = 0 To vmr.ValueCount - 1
    rv.Value(i) = vmr.Value(i)
    rv.symbol(i).Color = vmr.symbol(i).Color
    rv.symbol(i).Style = vmr.symbol(i).Style
```

```
    rv.symbol(i).Size = vmr.symbol(i).Size
   Next
   pathName = Left$(drawLayer.Tag, renderPos)
   curRenderName = "UniqueValue"
   drawLayer.Tag = pathName + curRenderName
  End If
 Case 3
  'Check if user clicked OK without settting a variable
  If cboQuantiles.List(cboQuantiles.ListIndex) <> "None" Then
   Dim rc As New MapObjects2.ClassBreaksRenderer
   Set drawLayer.Renderer = rc
   With cbr
    rc.Field = .Field
    rc.SymbolType = .SymbolType
    rc.BreakCount = .BreakCount
    rc.RampColors pctStart.BackColor, pctEnd.BackColor
    For i = 0 To .BreakCount - 1
     rc.Break(i) = .Break(i)
     rc.symbol(i).Style = .symbol(i).Style
     rc.symbol(i).Size = .symbol(i).Size
    Next
   End With
   pathName = Left$(drawLayer.Tag, renderPos)
   curRenderName = "Quantiles"
   drawLayer.Tag = pathName + curRenderName
  End If
 End Select
 Form1.Map1.Refresh
 Unload Me
End Sub
```

17 In all other subs in this module, replace *drawLayer.Tag* with *curRenderName*. The routines affected are *RestoreUniqueValueMap*, *RestoreQuantileMap*, and *cmdCancel_Click*.

Saving a selected set required quite a bit of coding. You had to create the *cmdSave* button and determine when it should be activated. A *cmdSave_Click* sub was developed to save the active layer's selected features. This required first creating an empty shape file using the *TableDesc* object and then adding features to it. To keep from overwriting an open shape file, the tag property for layers was constructed to store the path and file name of each layer. The changes in steps 12 through 15 are simply "housekeeping" changes to accommodate the new tag structure for layers you implemented to preclude the user overwriting an existing open shape file.

Selecting by Theme

CD-ROM NOTE: *The VB project in the* Chapter12_3 *directory on the companion CD-ROM starts here.*

In the current program, users can select features by clicking on them or by drawing a rectangle. Suppose you wanted to provide the ability to select features in a layer based on their relationship to features in another layer, much like the Select By Theme option in ArcView 3. The following describes the process for doing this.

1. Add a new command button to form *Form1*. Name the new button *cmdIntersect*, and give it the caption *Intersect* (see figure 12-3).

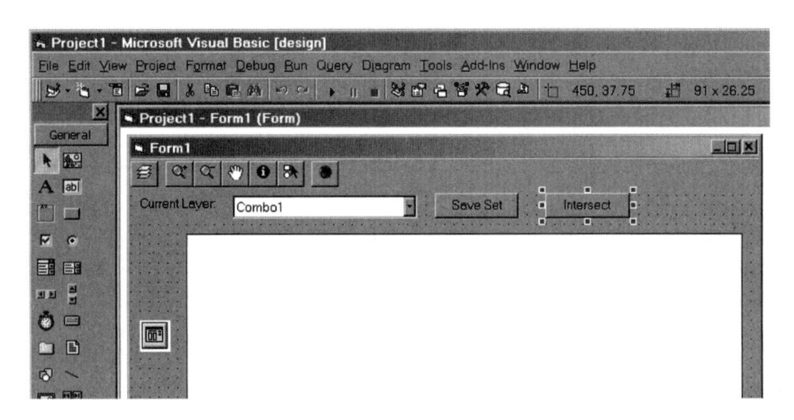

Fig. 12-3. Intersect button.

When should this button be enabled? Only when there are at least two layers in the map and there is an active layer. Thus, in *RefreshCombo1*, you need to add code that turns on the Intersect button when there are more than two layers and turns off the Intersect button otherwise.

2. Edit *RefreshCombo1* so that it reads as follows. Changes are indicated in bold.

```
Public Sub RefreshCombo1()
  Dim i As Integer
  Dim curselected As String
  If (Combo1.ListIndex >= 0) Then
    curselected = Combo1.List(Combo1.ListIndex)
    Toolbar1.Buttons("Identify").Enabled = True
    Toolbar1.Buttons("Select").Enabled = True
    cmdSave.Enabled = True
    If (Map1.Layers.Count > 1) Then
      cmdIntersect.Enabled = True
    Else
      cmdIntersect.Enabled = False
```

```
    End If
  Else
    curselected = ""
    Toolbar1.Buttons("Identify").Enabled = False
    Toolbar1.Buttons("Select").Enabled = False
    cmdSave.Enabled = False
    cmdIntersect.Enabled = False
  End If
  Combo1.Clear
  If Map1.Layers.Count = 0 Then
    Exit Sub
  End If
  For i = 0 To Map1.Layers.Count - 1
    Combo1.AddItem Map1.Layers(i).Name
    If Combo1.List(i) = curselected Then
      Combo1.ListIndex = i
      Set ActiveLayer = Map1.Layers(i)
    End If
  Next i
End Sub
```

You also need to enable the Intersect button when an active layer is set and the number of layers is two or more. That is, *Combo1_Click* becomes the following.

3 Enter the following lines at the end of the *Combo1_Click* sub. Changes to the previous version of the sub are indicated in bold.

```
Private Sub Combo1_Click()
 If ListIndex >= 0 Then
  Set ActiveLayer = Map1.Layers(Combo1.ListIndex)
  Toolbar1.Buttons("Identify").Enabled = True
  Toolbar1.Buttons("Select").Enabled = True
  cmdSave.Enabled = True
  If (Map1.Layers.Count > 1) Then
   cmdIntersect.Enabled = True
  End If
  'This works but we must keep the combo list and the
  'map layers synchronized
 End If
End Sub
```

When the user clicks on the Intersect button, the program should present a new form.

4 Create the following sub in form *Form1*'s code page.

```
Private Sub cmdIntersect_Click()
 frmIntersect.Show
End Sub
```

5 Add a new form to your project, and name it *frmIntersect.*

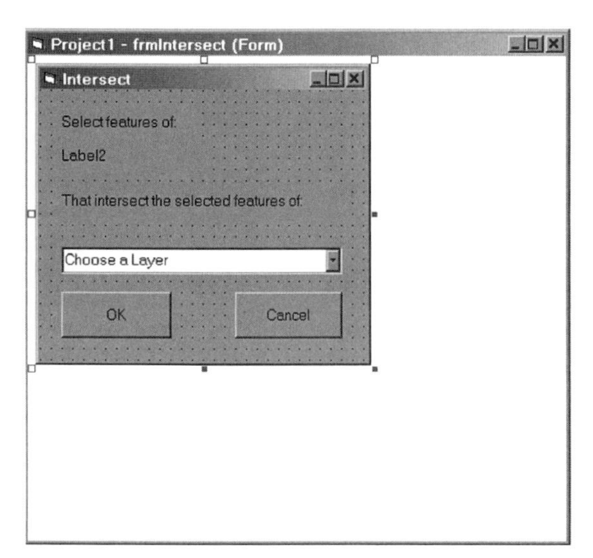

Fig. 12-4. The frmIntersect *form.*

Add the controls depicted in figure 12-4 (three labels, a combo box, and two command buttons).

Two of the labels you add to the form should not change: "Select features of:" and "That intersect the selected features of:". The label (*lblActiveLayer*) will display the name of the active layer, and the combo box (*cboLayers*) will list all layers except the active layer. Finally, the two command buttons are the usual OK and Cancel buttons. Give the combo box a default text value of *Choose a Layer.* When the form is called, the proper values are loaded into *lblActiveLayer* and *cboLayers* by the following code.

6 Create the *frmIntersect Load* sub to read as follows.

```
Private Sub Form_Load()
 Dim lyr As MapObjects2.MapLayer
 lblActiveLayer.Caption = ActiveLayer.Name
 For Each lyr In Form1.Map1.Layers
  If lyr.Name <> ActiveLayer.Name Then
   cboLayers.AddItem lyr.Name
  End If
 Next
End Sub
```

The only other things you need to specify for this form are the actions that correspond to the command buttons. The Cancel button merely unloads the form.

7 Enter the following.

```
Private Sub cmdCancel_Click()
 Unload Me
End Sub
```

The OK button needs to check if an intersect layer was chosen. If so, it will call a new sub (stored in form *Form1*'s code page) and pass it the name of the layer in the combo box.

8 Enter the following.

```
Private Sub cmdOK_Click()
 If cboLayers.ListIndex >= 0 Then
  Form1.Spatial_Select (cboLayers.List(cboLayers.ListIndex))
 End If
 Unload Me
End Sub
```

The heart of this operation is in the sub *Spatial_Select*, which you will add to form *Form1*'s code page. In many ways, this is similar to the case of the user selecting map features by clicking on an item or drawing a rectangle. However, the program needs to check the collection *colRecSetClass* for two layers: the active layer and the intersect layer. You will use *refIndex* to store the index number of the intersect layer in the *colRecSetClass* collection. For the active layer, this value will be called *curIndex*. Because this is a complex sub, its presentation is broken into steps 9 through 11.

9 Begin the *Spatial_Select* sub by adding the following sub to form *Form1*'s code page.

```
Public Sub Spatial_Select(lyrname As String)
  Dim curIndex, refIndex As Integer
  Dim aRec As MapObjects2.Recordset
  Dim i as Integer
  curIndex = -1
  refIndex = -1
```

The program next checks to see if there are any elements in the *colRecSetClass* collection. If not, it adds an element for the active layer and sets *curIndex* to 1. If there are elements in the collection, the program cycles through all elements to see if either the active layer or the intersect layer are in the collection. If after this process *curIndex* is still –1, the active layer is added to the *colRecSetClass* collection.

10 Enter the following.

```
  If (colRecSetClass.Count = 0) Then
     Dim bLayer As New clsRecSet
     bLayer.Name = ActiveLayer.Name
     colRecSetClass.Add bLayer
     curIndex = 1
    Else
```

```
  For i = 1 To colRecSetClass.Count
   If colRecSetClass.Item(i).Name = ActiveLayer.Name Then
    curIndex = i
   End If
   If colRecSetClass.Item(i).Name = lyrname Then
    refIndex = i
   End If
   If (curIndex >= 0) And (refIndex >= 0) Then
    Exit For
   End If
  Next i
 End If
 If (curIndex = -1) Then
  bLayer.Name = ActiveLayer.Name
  colRecSetClass.Add bLayer
  curIndex = colRecSetClass.Count
 End If
```

You are now ready to try the spatial selection. To implement this, you will use the *MapLayer* object's *SearchShape* method. *SearchShape* takes three arguments: a geometric shape or record set, a search method, and an SQL statement. For the current operation, the first argument will be a record set.

There are two possibilities to consider. If the intersect layer has no selected elements, you will use the entire intersect layer. If there are selected records in the intersect layer, you will use only those in *SearchShape*. The following is the remainder of the *Spatial_Select* sub.

11 Enter the following.

```
If refIndex = -1 Then
   Dim allrecs As New MapObjects2.Recordset
   Dim lyr As MapObjects2.MapLayer
   For Each lyr In Map1.Layers
    If lyr.Name = lyrname Then
     Set allrecs = lyr.Records
     Exit For
    End If
   Next
   Set gSelection = ActiveLayer.SearchShape(allrecs, _
       moAreaIntersect, "")
   gSelection.MoveFirst
   Do While Not gSelection.EOF
    Set aRec = ActiveLayer.SearchExpression("featureId = " _
        & gSelection.Fields("FeatureID").ValueAsString)
    colRecSetClass.Item(curIndex).RecSetCol.Add aRec
```

```
    gSelection.MoveNext
   Loop
  Else
   Dim currec As New MapObjects2.Recordset
   For i = 1 To colRecSetClass.Item(refIndex).RecSetCol.Count
    Set currec = colRecSetClass.Item(refIndex).RecSetCol.Item(i)
    Set gSelection = ActiveLayer.SearchShape(currec, _
          moAreaIntersect, "")
    If gSelection.Count > 0 Then
     gSelection.MoveFirst
     Do While Not gSelection.EOF
      Set aRec = ActiveLayer.SearchExpression("featureId = " & _
          gSelection.Fields("FeatureID").ValueAsString)
      colRecSetClass.Item(curIndex).RecSetCol.Add aRec
      gSelection.MoveNext
     Loop
    End If
   Next
  End If
  Map1.Refresh
End Sub
```

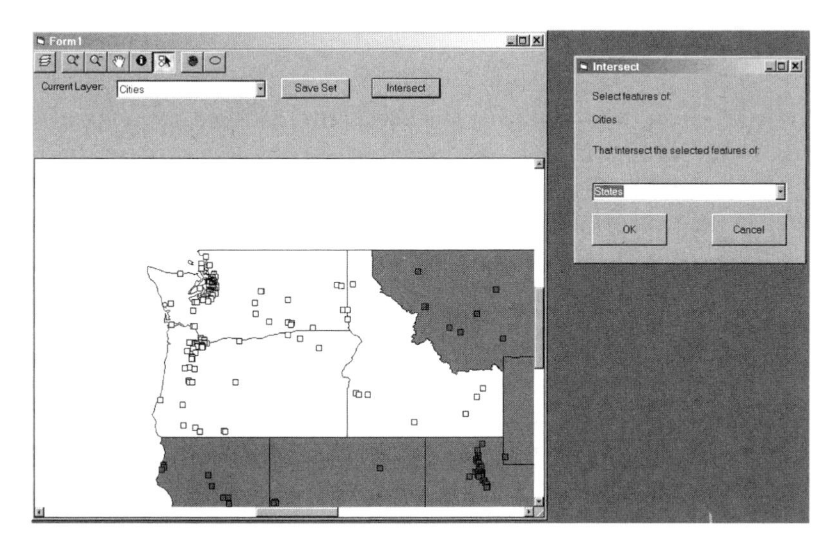

Fig. 12-5. Selecting cities in Washington, Oregon, and Idaho.

In the first case (*refIndex* = *-1*), there are no selected records in the intersect layer. Therefore, the program searches *allrecs*. In the second case, the program searches only for intersections of the selected features of the intersect layer. In either case, once there are new records (*gSelection.Count* is greater than 0), they are added one at a time to the *RecSetCol* of the *curIndex*'s *colRecSetClass*. If you run the program, you can create selections like those shown in figure 12-5. Adjunct exercise 12-2, which follows, provides you with the opportunity to practice adjusting selection parameters.

Adjunct Exercise 12-2: Fine Tuning Selection Parameters

This spatial select method works, but it is not without its problems. First, if the active layer already has some selected features, the *Spatial_Select* will add to the selected set. This may not be what you want. Try implementing the options New Set, Select From Set, and Add to Set. These can be check boxes on the *frmIntersect* form.

A second problem is that the current method may select features that are already selected. Try adding code, similar to that in the *MouseDown* event, to guard against this happening. Finally, the current version supports *moIntersect* only. There are many other ways to search shapes. Try incorporating options on the form *frmIntersect* to allow the user to select the Select by Theme option.

Summary

Creating selected sets, either by using the mouse or by spatial intersection, is a complex task. Saving those selections required you to store the path and file name for each layer in its *Tag* property. This was necessary to ensure that you did not try to overwrite an existing open shape. The numerous steps and changes in the code indicate the complex nature of the topics covered in this chapter.

There are many more aspects of interactive programming you could add to this program. However, let's move on to the topic of web-enabling the programs you have written. Web-based GIS is an important tool for disseminating geographic information to a large number of people. This is the topic of the next three chapters. As you will see, you can use much of the functionality you have already developed.

Chapter 13

Web Basics

Introduction

Web-based GIS is an exciting new area of application development. Although serving static maps has been around since the earliest days of the Internet, it is only recently that interactive maps have been available on the Web. Before you can consider using the program developed in the previous chapter on the Internet, you must have a basic understanding of how communication on the Web takes place. That is the purpose of this chapter.

GIS and the Web

You have probably seen various types of web-based mapping. For example, there are numerous sites that support address mapping. Many government agencies have sites that allow you to map specific data, such as Superfund sites or watersheds. Figure 13-1, for example, shows a map of solid waste disposal sites in Tennessee. In addition to the map display, this site supports zoom, pan, and identify tools, with digital images as part of the identify process. Many web sites support various GIS functions, and such sites can be based on very different design strategies.

NOTE: *This text does not consider serving static maps as a web-based GIS application, although static maps may be useful ways of conveying information. For purposes of this text, think of web-based GIS as those technologies that allow the user to interact with the map. Users should be able to determine the variables to be mapped and the method of thematic display, as well as pan, zoom in or out, and identify features on the map.*

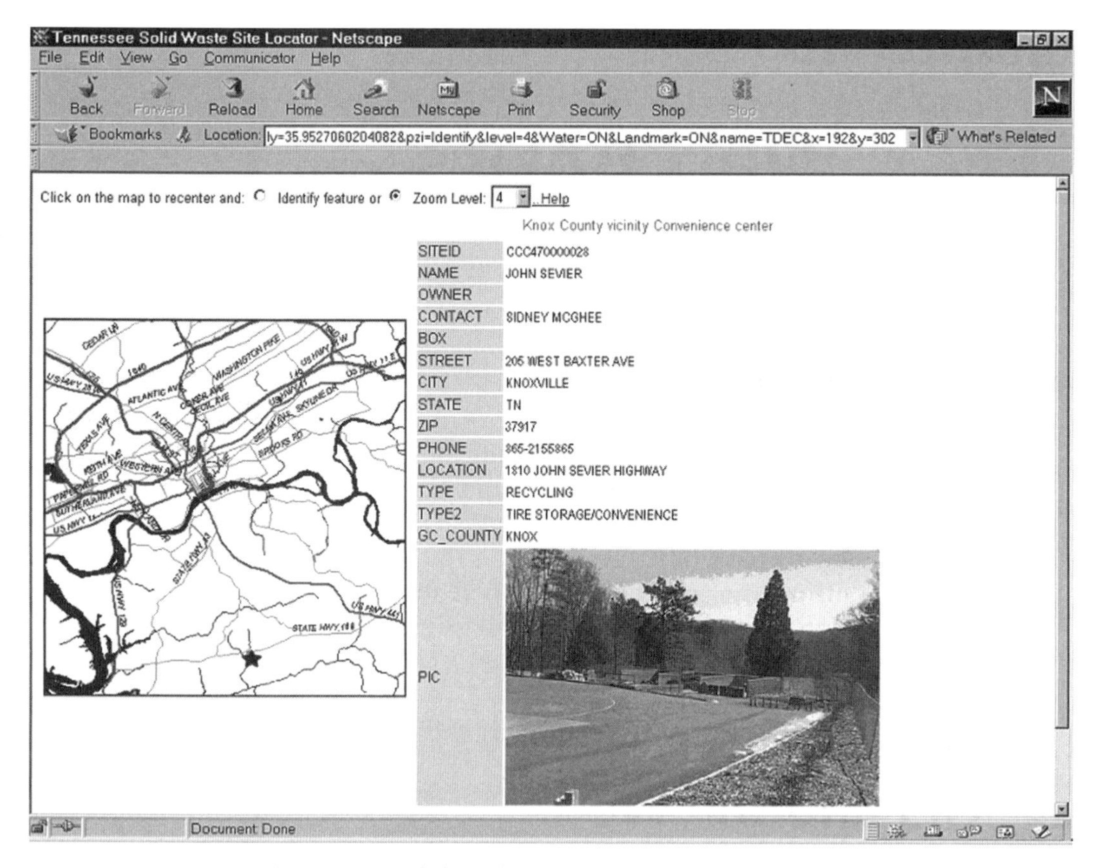

Fig. 13-1. A web-based GIS site.

Web GIS applications involve a user (the client), who contacts a server for some information. You might think of the spectrum of strategies used to implement GIS on the Web as involving two extremes. The first extreme is for the server to pass data and mapping applications (usually Java applets) to the client (you might call this the "server-supplied" strategy). That is, the server supplies the data and the programs, but all GIS functions are carried out on the client side (see figure 13-2).

The second strategy (the "client-requested" strategy) is to have the client indicate the type of map (or map function) she wishes to execute, and for the server to pass back the map the client requested (see figure 13-3).

You might ask which strategy is better for a given implementation. This depends on several factors. If you have visited a site that

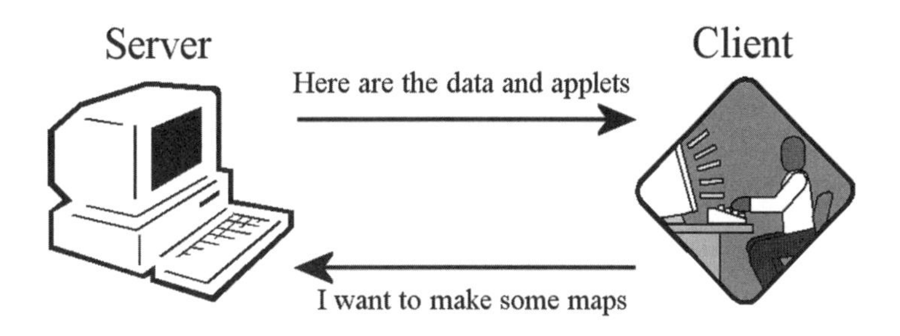

Fig. 13-2. The "server-supplied" strategy.

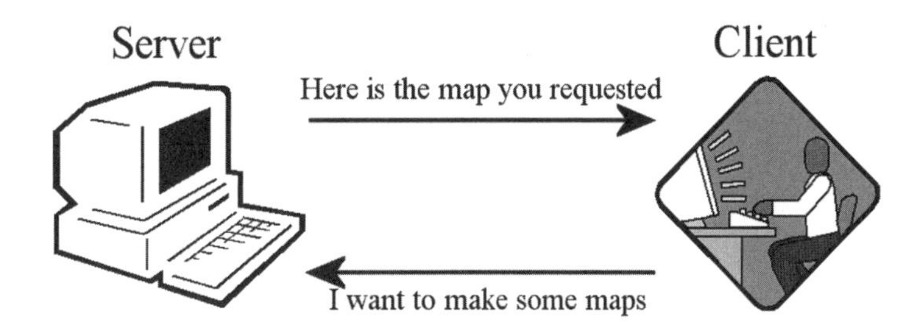

Fig. 13-3. The "client-requested" strategy.

passes data and Java applets (or some other program plug-in, such as ActiveX controls), chances are you were impressed with the functionality and ease of use. However, in this case it is a safe bet that the amount of data used in the mapping was quite modest. That is, this type of implementation is unlikely to involve downloading a shape containing 100,000 features, or, even more extreme, several huge shape files.

The reason for this is clear. Huge shapes will take a long time to transmit, and will require that the client have enough disk space and/or RAM to store them. Thus, the following becomes a rule: *Do not use the server-supplied strategy if your data sets are very large.* One might argue that Internet speeds will only increase, and storage is getting cheaper. These are valid points. Perhaps our rule has a limited life span. However, there is another situation in which you do not want to use the server-supplied strategy. Suppose you had sensitive data, such as company sales, patient records, or some other proprietary information. Do you want people downloading this raw data? The following is another valid rule: *Do not use the server-supplied strategy if you are concerned about data security.*

The client-requested strategy is the approach MO supports. It is not without its drawbacks. For example, if your site receives a lot of traffic, the server is going to be very busy processing each user's request for a map. Thus, you arrive at another rule: *Do not use the client-requested strategy if you are worried about server overload.*

There are some things, however, you can do to limit the stress on your server. For example, you can incorporate some data-checking and form-processing tasks in your web page that are to be processed by the client. In fact, as web-based scripting languages (such as Java Script and VB Script) get more powerful, you will find yourself using a hybrid of the two strategies. Remember, these strategies represent extremes. Some Internet map server software allows you to distribute the load over several servers.

If you are going to use MO to generate maps, and are going to display those maps on the Web, you must have a web server program on the server (such as Sambar, Apache, or Microsoft Internet Information Services). This is the program that handles the communication between the client and the server. You might think of it as a traffic cop that passes the requests of the client to the server programs.

If you want to serve maps using MO, you have two strategies. In the first, messages from clients are processed by an Internet map server (IMS) (such as the MapObjects IMS or ArcIMS). The IMS constitutes another layer of software that takes input from the server software, parses it, and passes it on to your VB/MO program. Among other operations, it then takes the VB/MO output and directs it (with the help of the server program) to the client. (IMS is explored in greater detail in Chapter 15.)

The second strategy is to write code that will execute much of the work otherwise performed by the IMS. This is not as difficult as it may seem, as you will see in Chapter 14. Developing your own code to process the interaction between the client and the server may seem like more work (and to a degree it is), but working with this code will help you understand more fully just what the IMS is doing.

Before you can implement either strategy, however, you must first understand how web pages work. That is what the remainder of this chapter is about. The material presented in this chapter is meant to be just enough to meet your needs for serving VB/MO applications on the Web. It is not meant to be a complete discussion of the Web or of HTML. Nor does this chapter attempt to cover frames, Java Script, VB Script, XML, or JSP. There are numerous books available that cover those topics.

HTML Basics

Web pages are based on a special language known as a HyperText Markup Language, or HTML. HTML is a language that web browsers, such as the one you wrote code for in Chapter 2, can understand. It is a fairly easy language to understand, although advanced web pages can be quite complex. The key to understanding HTML is understanding tags.

A tag is keyword placed within angle brackets, such as *<HTML>* or *<TABLE>*. Most, although not all, tags are in pairs. The first part of the pair is the keyword in angle brackets. The second part is the same, except that the keyword is preceded by a forward slash, as in *</HTML>*. The first part tells the browser that reads the web page that some special effect is to be applied; the balancing tag tells the browser that the special effect is over.

The following represents a valid, functional web page. You will be using this in the tutorial that follows.

1. Open your favorite text editor, such as Notepad, and enter the following. Save it with an *html* (or *htm*) extension.

```
<HTML>
<HEAD>
<TITLE>
My Web Page
</TITLE>
</HEAD>
<BODY>
This is a web page
</BODY>
</HTML>
```

If you look at this web page in a browser, figure 13-4 shows you what you will see.

Note that at the top of the browser is the page's title (*My Web Page*). The text simply includes the notation *This is a web page.* (Not a very interesting page, but it is a start.)

Let's take a closer look at the code. The page starts with the *<HTML>* tag and ends with the balancing *</HTML>* tag. Between these tags, the page is broken into two parts: the body and the head. The head, which starts with the *<HEAD>* tag and ends with

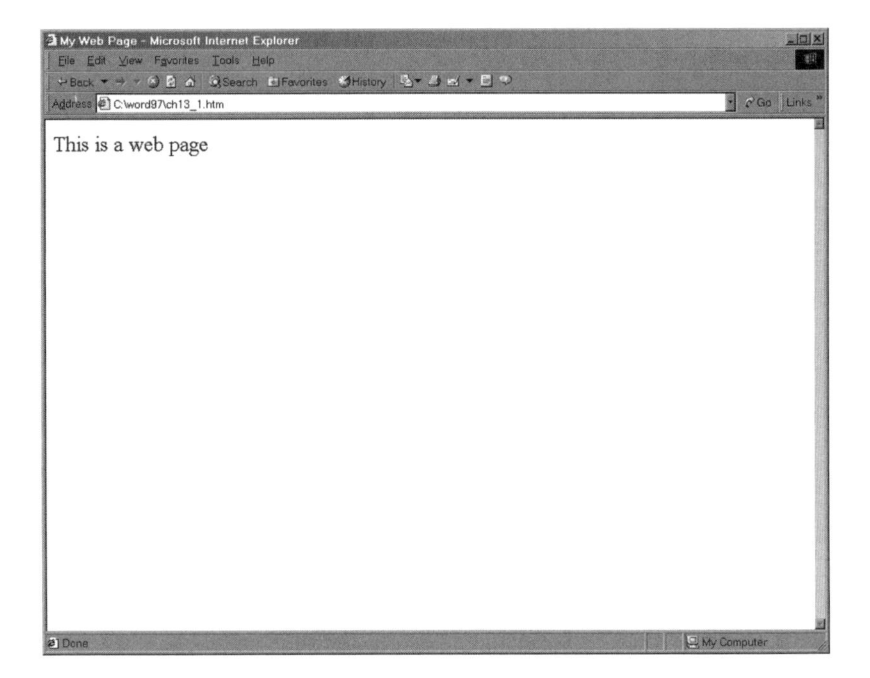

Fig. 13-4. First web page.

the *</HEAD>* tag, contains the title. The title of the page, which appears at the top of the browser, is enclosed by the *<TITLE>* and *</TITLE>* tags. The head can contain more information than the title, but for the purposes of this example, this is enough for now. Let's add some new lines to the page.

2 Return to your editor and edit the web page to read as follows (changes to the existing web page are indicated in bold).

```
<HTML>
<HEAD>
<TITLE>
My Web Page
</TITLE>
</HEAD>
<BODY>
<H1>The Great Programmer announces:</H1>
This is a web page
<P>This is another line of text.
This is another one on the same line.</P>
<P>Here is the line of decency</P>
<HR>
<B>This bold fellow is below the line of decency</B>
</BODY>
</HTML>
```

Figure 13-5 shows what the page looks like at this stage in a browser.

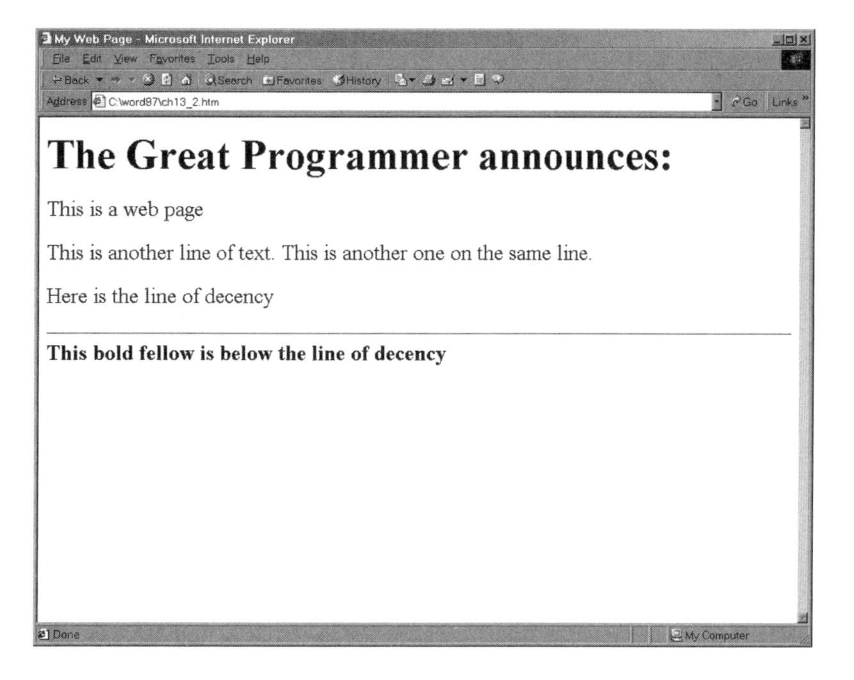

Fig. 13-5. Second web page.

The first new line contains the tags *<H1>* and *</H1>*, which indicate "heading, size 1." You have a choice of heading sizes, such as H2 or H3. H1 is the largest font. The next two lines of the previous code, as follows, are typed on separate lines.

```
<P>This is another line of text.
This is another one on the same line.</P>
```

However, on the web page these two lines appear on the same line. HTML does not recognize new lines as new lines. However, the tag *<P>* indicates a new paragraph, and hence a new line, and its balancing tag *</P>* indicates the end of the paragraph.

The tag *<HR>* is one of those tags that does not have a balancing, slashed tag. An HR tag places a line across the web page. In the previous code example, this tag line is labeled *<P>Here is the line of decency</P>*.

The following is the last new line in the previous code example.

```
<B>This bold fellow is below the line of decency</B>
```

The *<B>* tag indicates that the text that follows should be in bold type. In fact, all text will be in bold until the balancing *</B>* is encountered. If you wanted to render this text in italic instead, you would use the *<I>* and *</I>* tags. If you wanted bold and italics, you could combine them, as in the following.

```
<B>This bold fellow is <I>below</I> the line of
decency</B>
```

This would result in the following appearing on the web page.

```
This bold fellow is below the line of decency
```

Now that you have the flavor of HTML, let's examine one more way of presenting information on a web page: the table. The table, not surprisingly, is set off by a *<TABLE>* tag and a *</TABLE>* tag. A table consists of rows and columns. If you think back to the flex grid of Chapter 9, you started with an empty grid and added rows and columns as needed, filling in each cell as it was defined. You use a similar approach in regard to the table.

Suppose you wanted to present a table with four rows and two columns. The first row would contain the words *Variable* (in the first column) and *Value* (in the second column). (Does this look familiar? Look back at what you included in the identify form of Chapter 7.) In this example, there are three variables: *City, State,* and *Capital.* The following shows how you would build a web page that contains such a table.

3 Enter the following.

```
<HTML>
<HEAD>
<TITLE>
Table Page
</TITLE>
</HEAD>
<BODY>
<TABLE BORDER>
<TR>
 <TD><B>Variable</B></TD>
 <TD><B>Value</B></TD>
</TR>
<TR>
 <TD>City</TD>
 <TD>Knoxville</TD>
</TR>
```

```
<TR>
 <TD>State</TD>
 <TD>Tennessee</TD>
</TR>
<TR>
 <TD>Capital</TD>
 <TD>No</TD>
</TR>
</TABLE>
<P>Here's the same table without the border</P>
<TABLE>
<TR>
 <TD><B>Variable</B></TD>
 <TD><B>Value</B></TD>
</TR>
<TR>
 <TD>City</TD>
 <TD>Knoxville</TD>
</TR>
<TR>
 <TD>State</TD>
 <TD>Tennessee</TD>
</TR>
<TR>
 <TD>Capital</TD>
 <TD>No</TD>
</TR>
</TABLE>
</BODY>
</HTML>
```

Figure 13-6 shows the resulting page. Each table is set off by the table tag *<TABLE>*. However, the first table adds the word *BORDER* to its tag. The result is that the first table has a border drawn around all cells.

The next tag, *<TR>*, starts a row. This is followed by each column in the row. The columns are set off by *<TD>* and *</TD>* tags, one per column. After all columns (in this case, two) are defined, the row is ended with a *</TR>* tag. The next row starts with a new *<TR>* tag, and so on. Note that the bold tag is used *within* each column's definition.

Let's look at another important aspect of web pages: hyperlinks. Hyperlinks allow you to connect one page to another page, or part of a page to another part of the same page. Let's work with

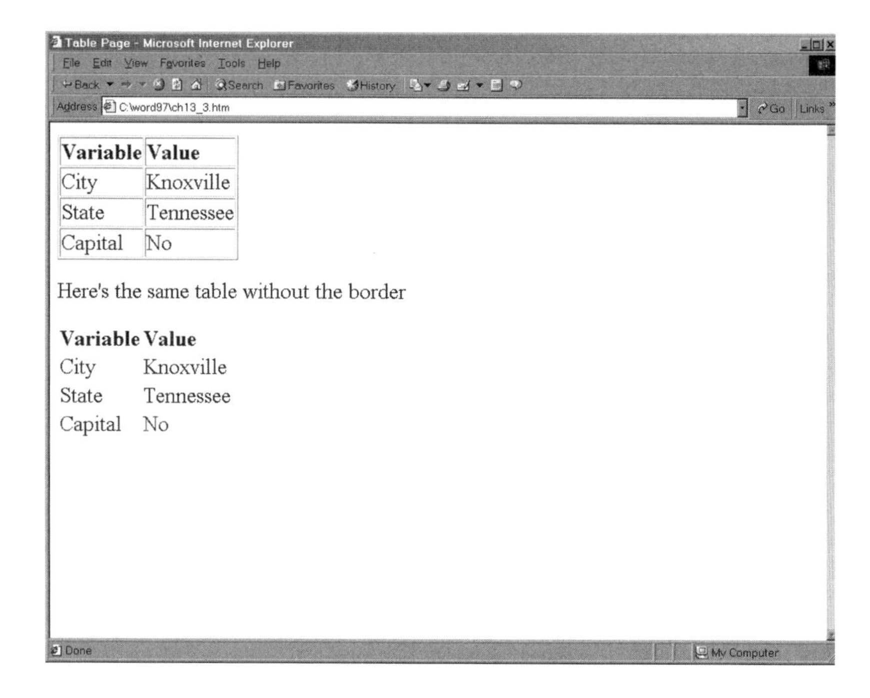

Fig. 13-6. Tables, with and without borders.

the first half of the previous *html* code. As in other instances, the only change to the previous code example, repeated in the following, is indicated by bold.

4 Make the following change (indicated in bold) to the previous code.

```
<HTML>
<HEAD>
<TITLE>
Table Page
</TITLE>
</HEAD>
<BODY>
<TABLE BORDER>
<TR>
 <TD><B>Variable</B></TD>
 <TD><B>Value</B></TD>
</TR>
<TR>
 <TD>City</TD>
 <TD><A HREF=http://www.utk.edu/maps/university/>Knoxville</A></TD>
</TR>
<TR>
 <TD>State</TD>
```

```
 <TD>Tennessee</TD>
</TR>
<TR>
 <TD>Capital</TD>
 <TD>No</TD>
</TR>
</TABLE>
</BODY>
</HTML>
```

This code inserts a hyperlink into the page. The tag *<A HREF=* indicates an action (hence, the *A*) to be taken when the user clicks on the hyperlink. The part following the equals sign to the closing angle bracket, *http://www.utk.edu/maps/university/>*, indicates the action to be taken. In this case, the action is to open a web page at the University of Tennessee (UT) that contains a map of the UT campus in Knoxville. The highlighted text in the web page exists between the closing angle bracket and the balancing action tag. That is, the highlighted text is Knoxville, which lies between the closing angle bracket (>) and the *</A>* tag, as in *>Knoxville</A>*. If you run this program and click on the hyperlink, you will arrive at a page that looks like that shown in figure 13-7.

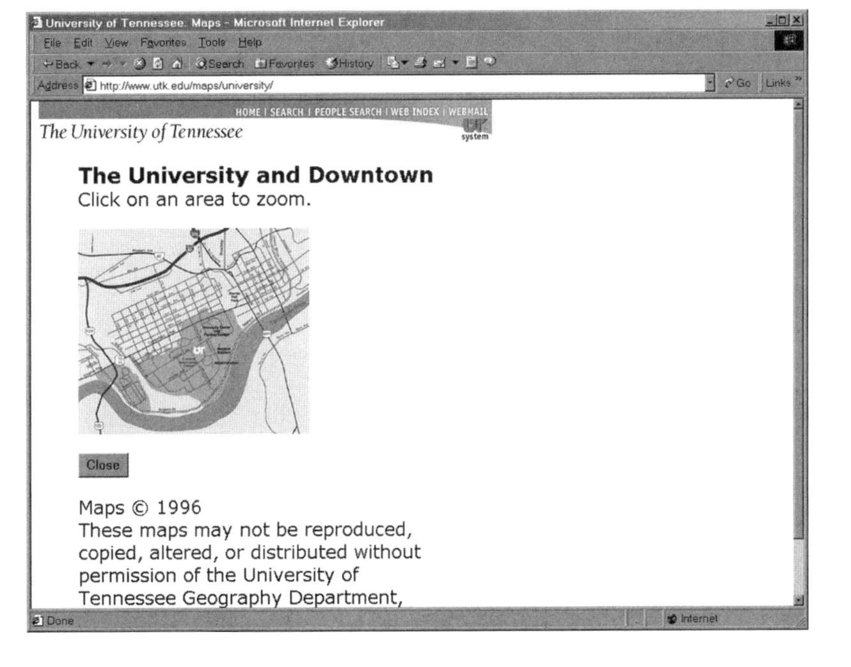

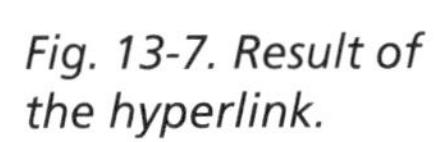
Fig. 13-7. Result of the hyperlink.

Can you see the table in this figure? It is the map. That is, the map is really a table consisting of three rows and three columns. Each cell in the map contains an image (a *gif* file). Each *gif* file (that is, each cell of the table) is a hyperlink. Put another way, a hyperlink can be either text or an image. The following is the section of code that places the map on the screen and sets the hyperlinks.

```
<table border=3 cellspacing=0 cellpadding=0>
<tr>
<td>
   <table border=0 cellspacing=0 cellpadding=0>
   <tr>
   <td>
   <a href="univ-northwest.html"><img width=98 height=85 border=0
      src="univ1-1.gif"></a><br>
   <a href="univ-west.html"><img width=98 height=85 border=0
      src="univ1-2.gif"></a><br>
   <a href="univ-southwest.html"><img width=98 height=85 border=0
      src="univ1-3.gif"></a></td>
   <td>
   <a href="univ-north.html"><img width=98 height=85 border=0
      src="univ1-4.gif"></a><br>
   <a href="univ-mid.html"><img width=98 height=85 border=0
      src="univ1-5.gif"></a><br>
   <a href="univ-south.html"><img width=98 height=85 border=0
      src="univ1-6.gif"></a></td>
   <td>
   <a href="univ-northeast.html"><img width=98 height=85 border=0
      src="univ1-7.gif"></a><br>
   <a href="univ-east.html"><img width=98 height=85 border=0
      src="univ1-8.gif"></a><br>
   <a href="univ-southeast.html"><img width=98 height=85 border=0
      src="univ1-9.gif"></a></td>
   </tr>
   </table>
</td>
</tr>
</table>
```

Study this *html* carefully. It illustrates how you can be very clever in constructing your web pages. What appears to be a map is a table with a border of width three, consisting of one cell. However, within that one cell is another table, consisting of three rows and three columns.

That table, the inner table, has no border. Each cell in it consists of an image, with a given height and width, in pixels, and no border. You know that these are images because they have *<img* tags.

The *<img* tag tells the browser that what follows will describe an image. In this case, the descriptors are height, width, border, and source (file name and location). Each image tag is surrounded by action tags (*<A HREF* and *</A>*). Thus, when a user clicks on any part of the map, the hyperlink is executed. For the UT map page, clicking on the map links the user with a more detailed map of the area on which the user clicked.

Tables, paragraphs, bold, italic, and hyperlinks present information to the user. However, what if you wanted to get information *from* the user? That is, how do you capture choices or information the user has made on a web page? To do this, you have to look at forms.

Form Basics

Forms are very powerful. They turn web readers from passive viewers into interactive participants. In the following you will learn how forms work, and will look at a few data entry methods available for working with forms.

Forms capture data and transmit it back to the server. How the server deals with this information is examined later in the chapter. For now, let's look at building forms.

Every form must have a *<FORM>* tag at the start of the form, usually after the *<BODY>* tag. A *<FORM>* tag must contain an *ACTION=* section, similar to a hyperlink, and a *METHOD=* section. One action you will make use of later in the chapter is to run a program that will echo back your inputs on a form. The *METHOD=* keyword indicates the method of communicating information from the web page to the program to be run. There are two methods: *POST* and *GET*.

How the programs on the server get the information from the client depends on the method of communication. The *POST* method uses a method called "standard in." You might think of "standard in" as information coming in as if it were typed from the keyboard. When you use *GET*, the information is passed in as command line arguments referred to as the query string. You have probably seen command line arguments; for example, the *RUN* dialog method for starting a program with a specific project. For example, the author types *excel.exe 1999.xls* on his home computer to display (in Microsoft Excel) students' grades for 1999. An example of how this is done is shown in figure 13-8.

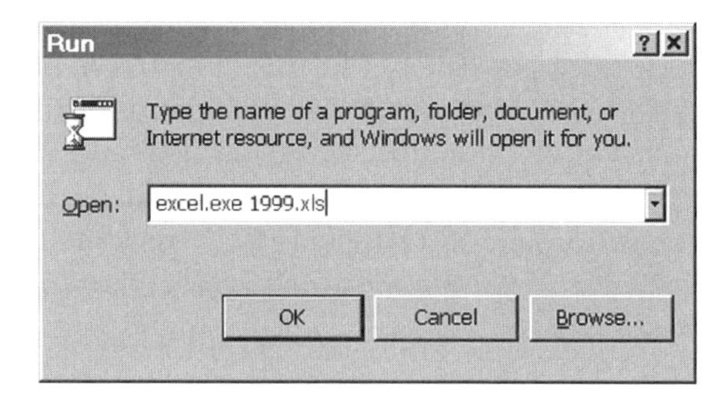

Fig. 13-8. Using a command line argument in the Run dialog.

Either method can be used. In the next chapter you will use the *POST* method. ESRI's MOIMS uses the *GET* method.

TIP: *Avoid passing very long strings of arguments with the GET method. They are likely to be truncated.*

You now know how to set up a form and have some sense of how it communicates with the server. In the next section, you will learn how to incorporate data input devices in web pages.

Web Page Data Input Devices

This section explores selection lists, click boxes, radio buttons, submit and image buttons, and hidden inputs. These are all types of data input devices. Let's look at the selection list. This is similar to list boxes in VB. In the following you will work with selection list.

1 Enter the following to create a web page.

```
<HTML>
<BODY>
<FORM ACTION="http://dellbar.geog.utk.edu/postprse.asp" METHOD = POST>
Select a render<BR>
<SELECT NAME="type">
 <OPTION VALUE="uniquevalue">Unique Value</OPTION>
 <OPTION SELECTED VALUE="classbreaks">Shaded or Class Breaks</OPTION>
 <OPTION VALUE="dotdensity">Dot Density<OPTION>
</SELECT>
<BR>
<INPUT TYPE=SUBMIT VALUE = "GO">
<INPUT TYPE=RESET>
</FORM>
</BODY>
</HTML>
```

The first two lines of this page are familiar. The third line tells the browser that this is a form, and that the program to which it responds is on the server at *dellbar.geog.utk.edu.* The name of the program is *postprse.asp.*

CD-ROM NOTE: *The program* postprse.asp *is found on the companion CD-ROM as is a PERL equivalent,* parseit.pl. *If you have a web server that supports active server pages* (asp)*, you can use* postprse.asp *on your own web site. If your server supports PERL, you can use* parseit.pl *on your web server. In either case, you will need to change the location of the server to reflect your server.*

In the previous code, the method for transmitting information from the client to the server is *POST.* The program then places a prompt on the screen (select a renderer).

The next portion of the code represents the selection box. A selection box is a particular type of input method. You define a selection box with the *<SELECT NAME=* tag. The *NAME=* component of this tag is crucial. It associates a variable name with the input method. When the information is sent to the server, it is sent in pairs consisting of variable name and value. Each option in the selection list appears between the *<SELECT NAME=type>* tag and the balancing *</SELECT>* tag. Each option prompt is surrounded by *<OPTION VALUE=>* and *</OPTION>* tags, in which *prompt* is the value the user sees. Let's look at one of these in more detail, such as the following.

```
<OPTION VALUE="uniquevalue">Unique Value</OPTION>
```

In this example, the user will see the string *Unique Value* in the list. If that string is chosen, the value of the input variable *type* is set to *uniquevalue.* The keyword *SELECTED* in an option indicates that this is the default value. The last two new lines in the previous code example, repeated in the following, incorporate Submit and Reset buttons.

```
<INPUT TYPE=SUBMIT VALUE = "GO">
<INPUT TYPE=RESET>
```

Every form must have a Submit button (or Submit image, discussed in material to follow), and should have a Reset button. The Submit button initiates a call from the client to the server to run the program listed to the right of *<FORM ACTION=.* The Reset button clears all inputs and resets them to their default values.

Suppose a user runs this program, selects Unique Value, and then clicks on the Submit button. Figure 13-9 shows the web page the server will send back to the client.

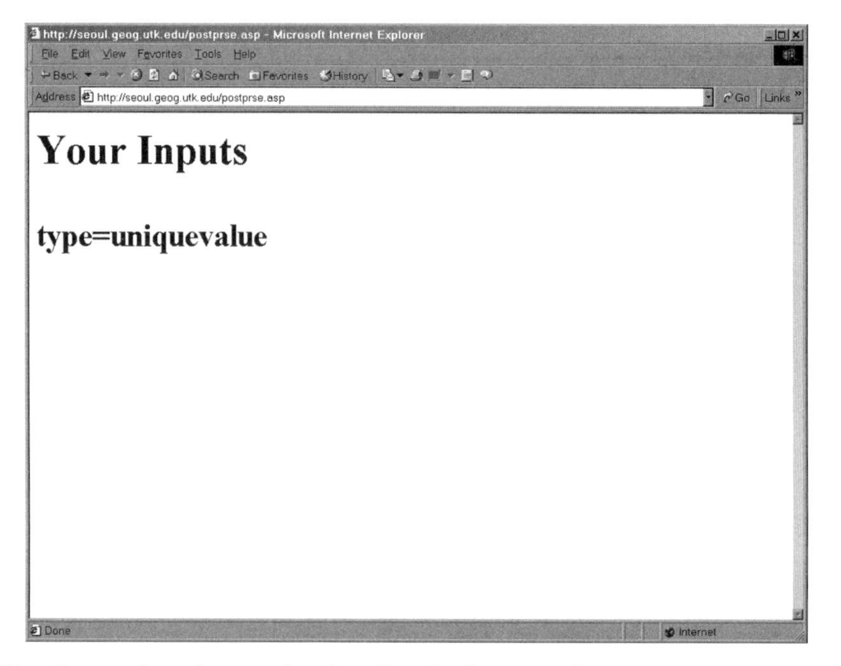

Fig. 13-9. Web page created by postprse.asp.

One last point about selection lists before moving on: you can set the number of rows visible in the list. By default, the value of the number of rows is 1, making the selection list a drop-down list. You could have incorporated a height value, using the *SIZE* keyword, when you declared the select list. Try using the following line at the start of the select list.

2 Enter the following.

```
<SELECT NAME="type" SIZE=4>
```

The second type of input to consider is the check box. Check boxes are relatively easy to incorporate. You simply declare an input to be a check box, indicate the corresponding variable name, and then list the prompt. If you want to have a box checked by default, you can add the *CHECKED* keyword to its declaration.

3 Add the following to your page directly after the *</SELECT>* tag, and rerun the program.

```
<BR>
Indicate layers to include<BR>
<INPUT TYPE=CHECKBOX NAME="state" CHECKED>State<BR>
<INPUT TYPE=CHECKBOX NAME="city">City<BR>
```

```
<INPUT TYPE=CHECKBOX NAME="roads">Roads<BR>
<INPUT TYPE=SUBMIT VALUE = "GO">
<INPUT TYPE=RESET>
```

Figure 13-10 shows what the page looks like after step 3.

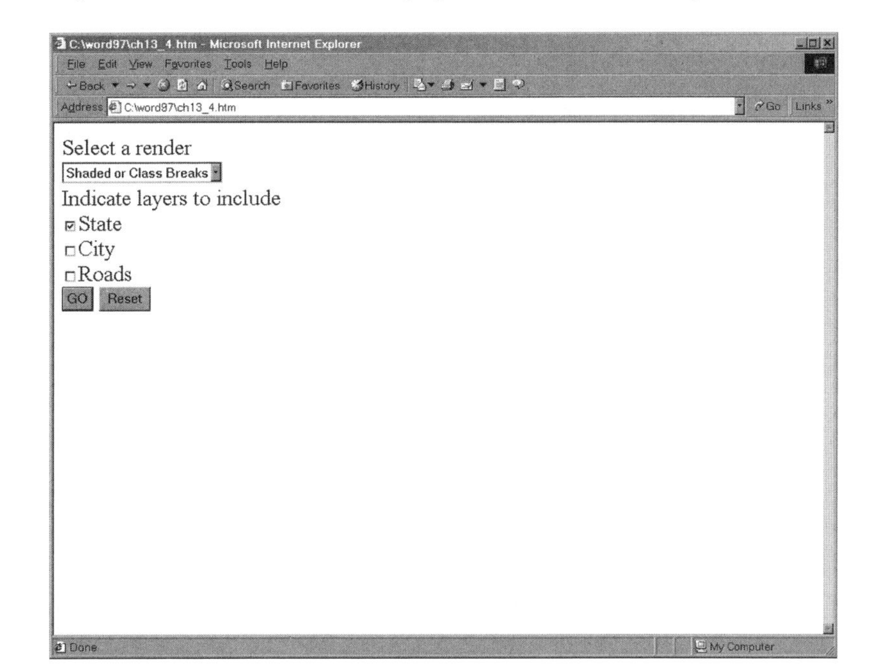

Fig. 13-10. Page with a Select list and check boxes.

Figure 13-11 shows what the user's pressing of the GO button will produce (using the defaults).

Notice what happens with check boxes. If a check box is not chosen, no value for it is sent from the client to the server. If it is checked, a value of "on" is returned. This makes check boxes function as Boolean variables.

The next input to consider is the radio button. The structure of a radio button's tags is similar to that for a check box. However, all options that are mutually exclusive must have the same name.

4 Add the following to your page, immediately after the code for the check boxes.

```
Choose an Action:
<INPUT TYPE=RADIO NAME="pzi" VALUE="pan">Pan<BR>
<INPUT TYPE=RADIO NAME="pzi" VALUE="in" CHECKED>Zoom In<BR>
<INPUT TYPE=RADIO NAME="pzi" VALUE="out">Zoom Out<BR>
<INPUT TYPE=RADIO NAME="pzi" VALUE="identify">Identify<BR>
```

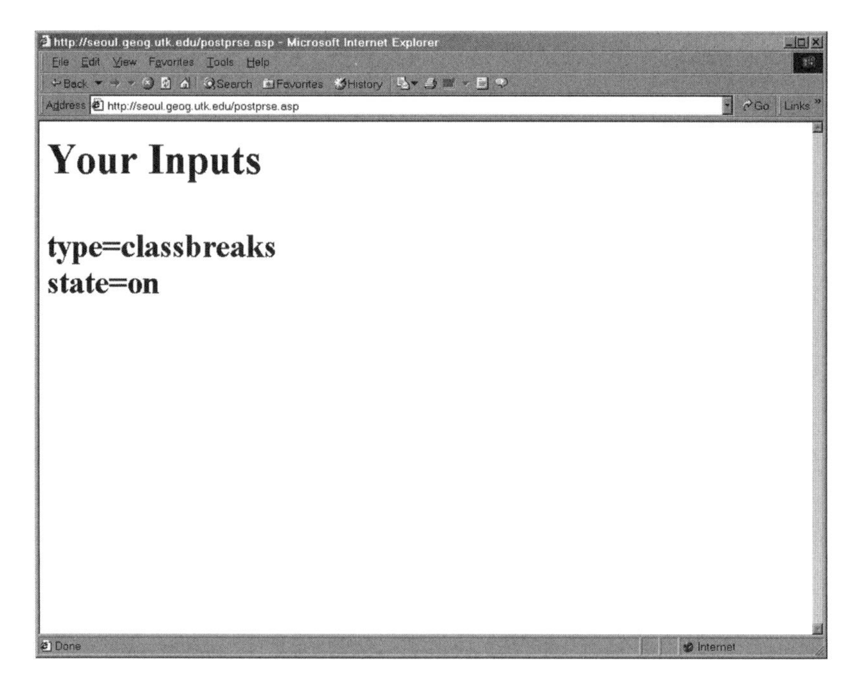

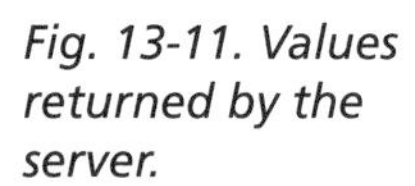
Fig. 13-11. Values returned by the server.

If you run this code, your page should look like that shown in figure 13-12.

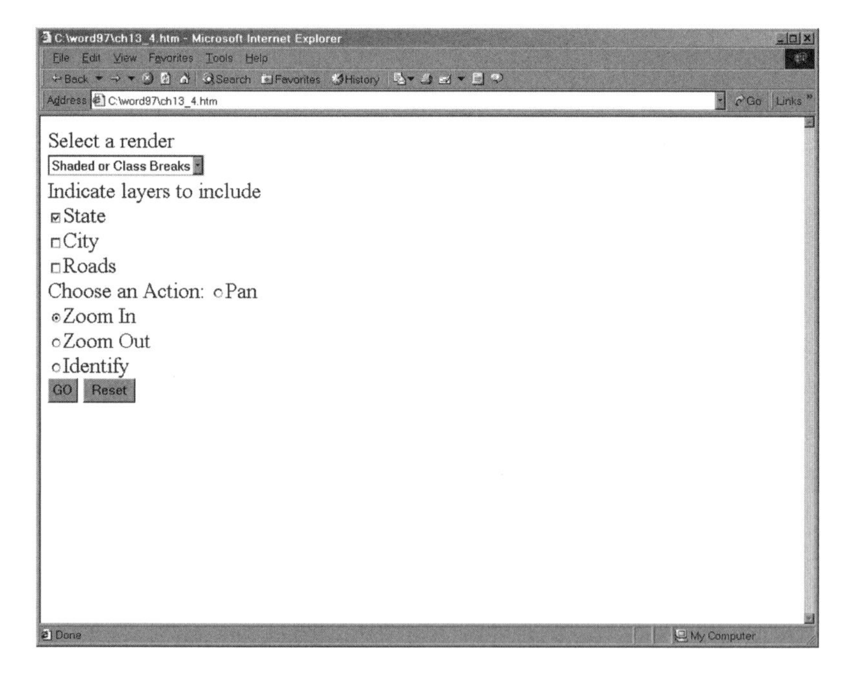

Fig. 13-12. New page with radio buttons.

Choosing some values and pressing the GO button will return something like that shown in figure 13-13.

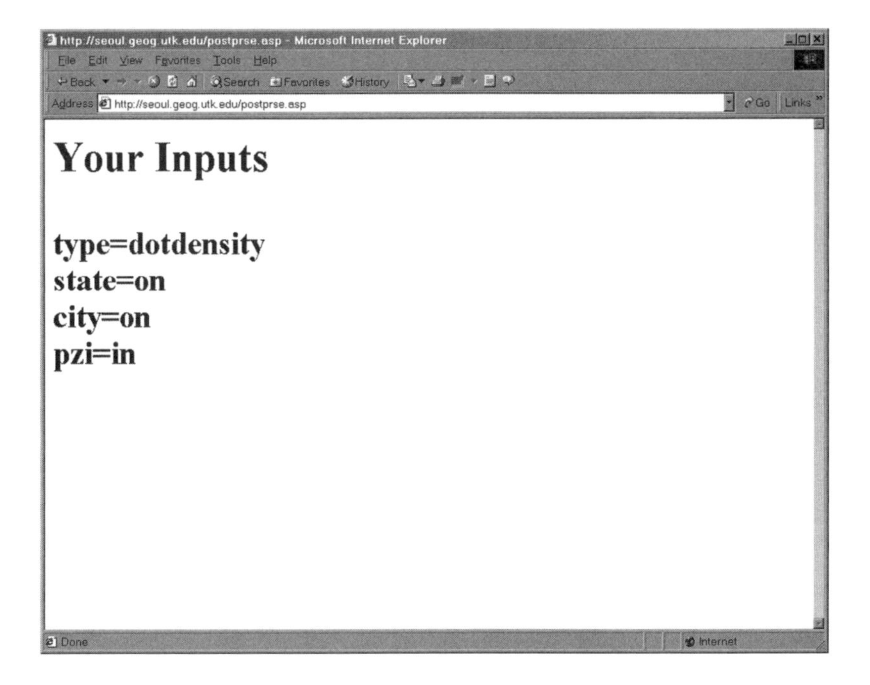

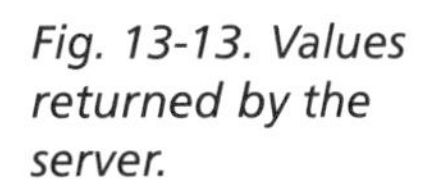
Fig. 13-13. Values returned by the server.

So far in this web page, you have initiated the action program by clicking on a Submit button. Alternatively, you can use images to initiate submits. To do this, you create an *IMAGE* input type and specify the name of the input file, as in the following.

```
<INPUT TYPE=IMAGE SRC="mymap.jpg">
```

In this example, when the user clicks on the image, the action listed in the form declaration takes place, just as with the Submit button. However, something more takes place. Two values, X and Y, are sent to the server. These indicate the location on the image, in pixels, where the user clicked. (Do you see how you might use this in an Identify button? A Zoom button?). The following is an example of this functionality in HTML code (the new line is in bold). Note the missing *SUBMIT* button.

5 Make the following code change (indicated in bold).

```
<HTML>
<BODY>
<FORM ACTION="http://dellbar.geog.utk.edu/postprse.asp" METHOD = _
   POST>
Select a render<BR>
<SELECT NAME="type" SIZE=4>
 <OPTION VALUE="uniquevalue">Unique Value</OPTION>
```

```
 <OPTION SELECTED VALUE="classbreaks">Shaded or Class Breaks</OPTION>
 <OPTION VALUE="dotdensity">Dot Density<OPTION>
</SELECT>
<BR>
Indicate layers to include<BR>
<INPUT TYPE=CHECKBOX NAME="state" CHECKED>State<BR>
<INPUT TYPE=CHECKBOX NAME="city">City<BR>
<INPUT TYPE=CHECKBOX NAME="roads">Roads<BR>
Choose an action<BR>
<INPUT TYPE=RADIO NAME="pzi" VALUE="pan">Pan<BR>
<INPUT TYPE=RADIO NAME="pzi" VALUE="in" CHECKED>Zoom In<BR>
<INPUT TYPE=RADIO NAME="pzi" VALUE="out">Zoom Out<BR>
<INPUT TYPE=RADIO NAME="pzi" VALUE="identify">Identify<BR>
<INPUT TYPE=IMAGE SRC="tenn.jpg">
<INPUT TYPE=RESET>
</FORM>
</BODY>
</HTML>
```

Figure 13-14 shows the web page after implementing this code.

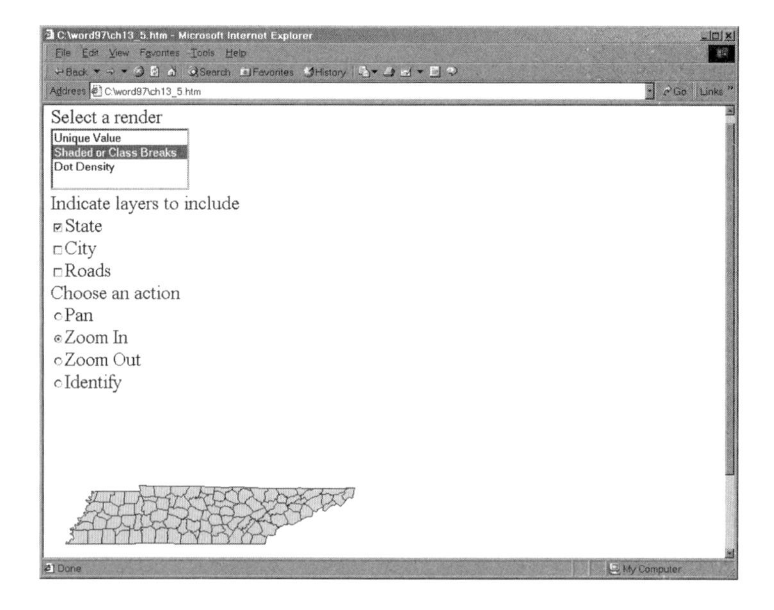

Fig. 13-14. Web page with an Image button.

Figure 13-15 shows the return from the server. Adjunct exercise 13-1, which follows, provides you with an opportunity to practice examining coordinate location.

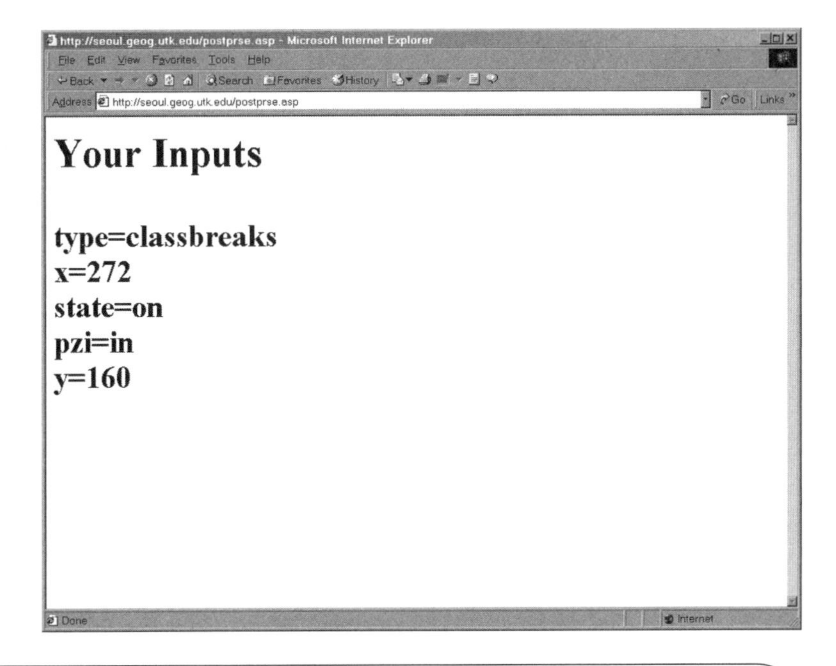

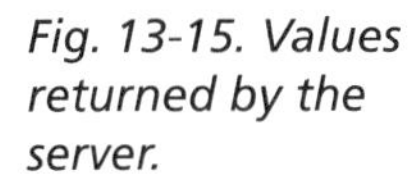
Fig. 13-15. Values returned by the server.

Adjunct Exercise 13-1: Examining Coordinate Location

Try clicking on various locationss on the map. Figure out where 0,0 is located and in what directions X and Y increase.

There is one last input type to be discussed: Hidden. You have probably filled out forms on the Web that consisted of more than one page. How does the final page know what has been selected before it? One method is to have hidden input; that is, input the user does not see, but the server will see. Suppose, for example, a user selected a variable to map on one page and a method of mapping on another. In this case, you would pass the following to the second page.

```
<INPUT TYPE=HIDDEN NAME="mapvar1" VALUE="pop1990">
```

In order to pass a hidden input to a web page based on user inputs from a pervious page requires that the second page be constructed after the user has selected the variable to be passed. For example, the previous line of code could not be written until the user selected the value *"pop1990"* from a previous web page.

Summary

This chapter began with a discussion of two primary strategies for incorporating GIS applications on the Web. To use MO and VB, the second strategy discussed works best. This strategy involves getting inputs from the client and pushing maps and web pages from the server to the client. You then examined basic text formatting in HTML, including how to build tables and add hyperlinks. All HTML features require the use of tags.

This chapter also explored the use of forms, which make the Web dynamic. Forms, it was discussed, require an action that will reside on the server. The information on the form can be passed from the client to the server as standard input (using the *POST* method) or as command line arguments (using the *GET* method). Forms can have many input types. Among those input types examined in this chapter were selection lists, check boxes, radio buttons, and image buttons.

You saw that with *POST* (as with *GET* in Chapter 15), each input is sent to the server paired with its value. The exception is check boxes, where only the names of the boxes checked along with the keyword *ON* are sent to the server. The last example in this chapter also showed that image buttons record the X and Y coordinates of a mouse click. It is this feature that allows you to implement features such as pan, zoom, and identify in Web-based GIS.

Chapter 14

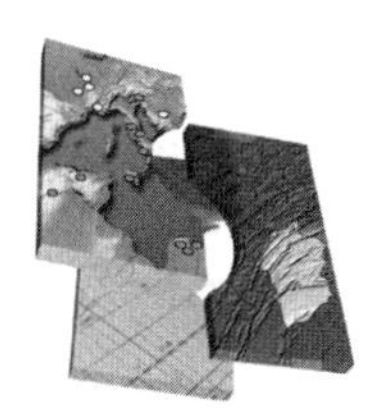

Serving Maps on the Web: Method 1

Introduction

Chapter 13 presented the basics of building web pages, particularly forms. You saw that data could be sent from a client to a server by one of two methods: *Post* or *Get*. In this chapter, you will combine that knowledge with the mapping program you created in previous chapters to serve interactive maps on the Web. All of the map processing will be done on the server, but the requests for functions such as zoom, pan, and identify will come from the client. This chapter also introduces using PERL programs to execute your VB/MO program on the server.

Overview of the Method

In this chapter you will use some of the methods discussed in Chapter 13 to serve maps on the Web. The following is an outline of the approach you will use in this chapter. (In Chapter 15 you will explore a second approach.)

- ❒ The user fills out a form requesting a particular map.
- ❒ When the form is submitted, the inputs are sent to the server. A PERL script (supplied on the companion CD-ROM for this tutorial) will then:
  - Parse the inputs and create a file of inputs and values
  - Call the VB/MO program, passing it the name of the file it just created as a command line argument
  - Suspend operation for a few seconds

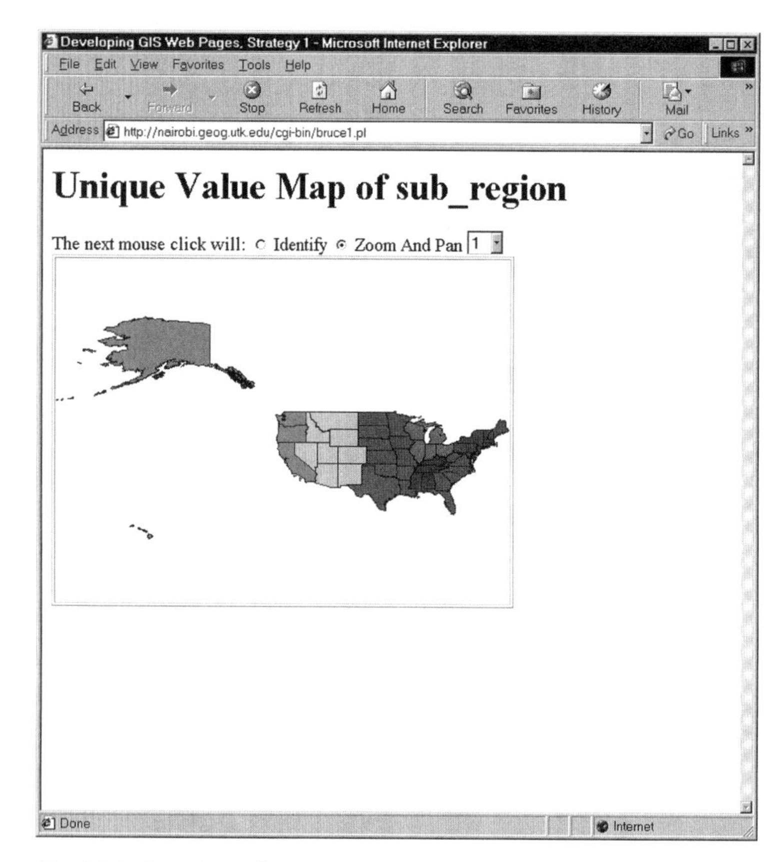

Fig. 14-1. Sample web page.

- The VB/MO program creates the requested map. After the map is created, the program calls a DLL (supplied on the companion CD-ROM for this tutorial) to convert the map to a JPG file.
- The PERL script then sends a web page back to the client. Embedded in the page are the Map, Zoom/Pan, and Identify buttons, as well as a set of hidden inputs (see figure 14-1).

The Tutorials

The map will be an input image, meaning that if the user clicks on the map the proper action will take place (discussed further in material to follow). For this chapter, assume that there can be at most three layers active on in the map: a *states* layer, a *cities* layer, and a *roads* layer. These are all found in the *shapes\USA* directory of the companion CD-ROM. You can use other shapes and more layers, but you will have to make some changes to the code that follows.

You will start by working with VB and the input file, simulating web operations. By the end of this chapter, you will hook everything together so that you can serve maps on the Web. To completely finish this chapter, you will need a web server computer that uses server software that supports PERL. (PERL is discussed at the end of the chapter.) Two options are Sambar and Apache.

Sambar is shareware, whereas Apache is freeware. You can read about these products at *http://www.sambar.com* and *http://httpd.apache.org*. Other packages are available.

1 Open a text editor and create a file with the following content.

```
00000
variable
pop1990
method
quantiles
classes
4
start
Yellow
end
Red
```

2 Save this file as *test.txt*.

You will use this file (and versions of it) to test your VB program before connecting to the Web. This test file will contain simulated inputs from a web page (which you will build in the following) for creating maps.

Modifying Form Units: From Twips to Pixels

CD-ROM NOTE: *The VB project in the* Chapter14_1 *directory on the companion CD-ROM starts here.*

In Chapter 13, you saw that if a user clicked on an input image, the web browser returned the *x,y* coordinates. These are in pixels. You therefore need to make sure your VB/MO project works in pixels. To do this, perform the following.

1 Open your project and bring up form *Form1*.
2 Set form *Form1*'s *ScaleMode* property to Pixel.
3 Select the map control and set its height and width to 400 pixels each. You may have to resize the form to hold the map control (see figure 14-2).

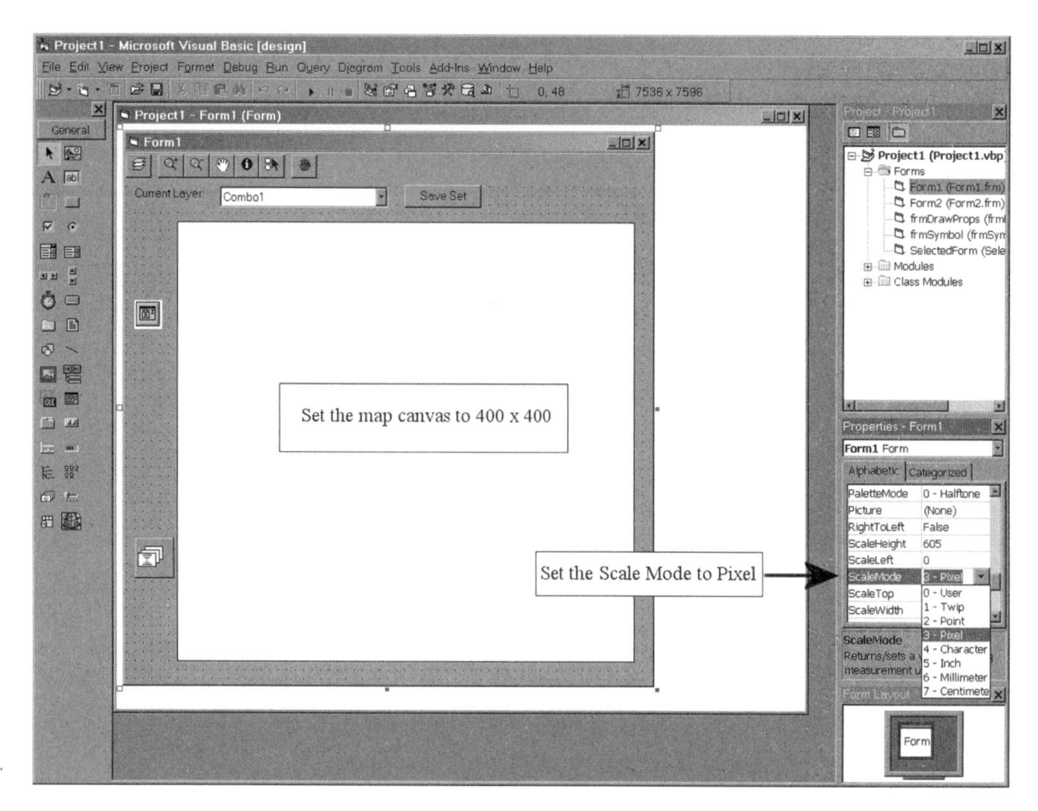

Fig. 14-2. Setting the ScaleMode *property and resizing the map.*

Getting Input File Values

The next thing you need to do is edit the *Form_Load* sub for form *Form1* to enable batch (noninteractive) mapping.

1 Edit the *Form_Load* sub as follows (changes are indicated in bold).

```
Private Sub Form_Load()
 RefreshCombo1
 Set gSelection = Nothing
 GetMappingValues
 If batchOn Then
MakeBatchMap
Unload Me
 End If
End Sub
```

The new sub, *GetMappingValues*, is where the program reads any input file that gets passed in as a command line argument. The Boolean variable *batchOn* lets the program know if it is doing batch mapping or interactive mapping. In other words, the VB/MO program will work as both an interactive program and batch (for web application) program. If in batch mode, the program will create the map (*MakeBatchMap*), and then end the program (*Unload Me*). Otherwise, the program will run as before. Before writing the complete *GetMappingValues* sub, you need to declare the following variables at the beginning of form *Form1*.

2 Declare the following additional variables (additions are indicated in bold).

```
Public colRecSetClass As New Collection
Dim batchOn As Boolean
Dim roadsOn, citiesOn As Boolean
Dim mapMethod As String
Dim mapVar As String 'quantiles and unique
Dim startClr, endClr, mapClasses As String 'quantiles
Dim mapClr, mapStyle As String 'single symbol or unique (mapStyle)
Dim mapPzi As String
Dim mapLevel As String
Dim minX, minY, maxX, maxY As String
Dim curX, curY As String
Dim jobID As String
```

The first Boolean, *batchOn*, indicates whether the program is running in interactive mode or batch mode. The next two Boolean variables, *roadsOn* and *citiesOn*, set the visibility of the *roads* and *cities* layers. The type of map to be created (single symbol, unique-value map, or quantile map) is stored in *mapMethod*. The variable used for quantile and unique-value maps is stored as *mapVar*. If a quantile map is being created, the program needs to keep track of the ramp colors (starting and ending) and the number of classes. This is what the next three variables do (*startClr*, *endClr*, and *mapClasses*).

If a single-symbol map is being created, the color is stored in *mapClr* and the shading style is stored in *mapStyle*. The latter variable is also used by the unique-value map renderer. Once the map is served to the client, the user can choose to either zoom/pan the map (zoom and pan are in the same command) or identify a state. In the VB program, that choice is stored in the variable *mapPzi*. If the user selects Zoom/Pan, the program needs to know the next level of zooming. That is stored in *mapLevel*.

The next four variables (*minX, minY, maxX,* and *maxY*) store the current extent of the map (in map units, which in this case is latitude/longitude), and *curX* and *curY* store where the user clicked on the map (in pixels). Finally, *jobID* stores the ID of the current job. Thus, when a user makes a request, he is assigned a *jobid.* This way, one client's map is not confused with another's. (Actually, there is a very small chance this could happen. Nevertheless, a solution is explored in Chapter 15.) Now you can create the *GetMappingValues* sub.

3 Enter the following.

```
Private Sub GetMappingValues()
 Dim infile As String 'file to open
 Dim inline As String 'line read from file
 Dim tempstr as String 'a throwaway string
 batchOn = False
 infile = Command()
 If infile = "" Then
  Exit Sub
 End If
 'If we get here, what do we know?
 batchOn = True
 mapLevel = "1"
 curX = "0"
 curY = "0"
 minX = "0"
 minY = "0"
 maxX = "0"
 maxY = "0"
 mapPzi = "zoom"
 mapLevel = "1"
 roadsOn = False
 citiesOn = False
 Open infile For Input As #1
 Line Input #1, jobID
 Do While Not EOF(1)
  Line Input #1, inline
  Select Case inline
Case "pzi"
    Line Input #1, mapPzi
   Case "level"
    Line Input #1, mapLevel
   Case "minx"
    Line Input #1, minX
   Case "miny"
```

```
    Line Input #1, minY
   Case "maxx"
    Line Input #1, maxX
   Case "maxy"
    Line Input #1, maxY
   Case "roads"
    Line Input #1, tempstr 'Throw away the word "on"
    roadsOn = True
   Case "cities"
    Line Input #1, tempstr
    citiesOn = True
   Case "method"
    Line Input #1, mapMethod
   Case "x"
    Line Input #1, curX
   Case "y"
    Line Input #1, curY
  'quantile variables
   Case "classes"
    Line Input #1, mapClasses
   Case "start"
    Line Input #1, startClr
   Case "end"
    Line Input #1, endClr
   'single value map
   Case "color"
    Line Input #1, mapClr
   Case "style"
    Line Input #1, mapStyle
   'quantiles or unique value
   Case "variable"
    Line Input #1, mapVar
  End Select
  Loop
End Sub
```

This code begins by declaring some necessary local variables. Next, *batchOn* is set to False, meaning that the default is for interactive mapping. The program then gets the command line argument (see Chapter 13) within *file = Command()*.

If there is a command line argument, it will be the name of the input file. If there is no file, the program exits the sub with *batchOn* as False, and the program runs as an interactive mapping program. If there is a command line argument, *batchOn* is set to

True and all variables that may not be in the input file are initialized to their default values.

After the initialization, the program loops through the *infile*, line by line. The program reads a line to get the type of variable being returned by the user, and then based on the case, reads the variable value. By the time the program exits the sub, it has read the entire input file. (The program reads the word *on* for check boxes to a temporary variable, *tempstr*, which it ignores.) Once you have these values, you can create the map. However, you first need to build two utility functions.

Translation Functions

When the user selects a color, the web page should return the selection. For example, if the user clicks on the word *Blue*, the web page should return the string *"Blue"*. In computers, colors are represented as numbers, not strings. Therefore, for example, you need to translate the word *Blue* into the proper number. You might wonder why you could not simply code the proper value into the web page. After all, if the color is selected from a list, the structure of the list prompt in HTML is *<OPTION VALUE=value>Prompt</OPTION>*.

You could set the prompt to *Blue* and the value to the appropriate value. That might work, but it is best to keep all MO translations in one place; namely, your VB/MO program. This makes maintenance easier—of both your program and your web page.

If you look up the MO Help for color constants, you will see that there are 23 values listed. The color constants have a name associated with them, such as *moBlue* for blue. The value of *moBlue* is 0xFF0000, a hex number. You need to translate the returns from the web page, which are words such as *Blue* or *Red*, into their respective numbers. The following function does this.

1 Add the following to the *modUtility.bas* code page.

```
Public Function LookUpColor(ByVal color As String) As Long
 Select Case color
 Case "Black":  LookUpColor = moBlack
 Case "Red":   LookUpColor = moRed
 Case "Green":  LookUpColor = moGreen
 Case "Blue":   LookUpColor = moBlue
```

```
 Case "Magenta":  LookUpColor = moMagenta
 Case "Cyan":   LookUpColor = moCyan
 Case "White":  LookUpColor = moWhite
 Case "Light Gray": LookUpColor = moLightGray
 Case "Dark Gray": LookUpColor = moDarkGray
 Case "Gray":   LookUpColor = moGray
 Case "Pale Yellow": LookUpColor = moPaleYellow
 Case "Light Yellow": LookUpColor = moLightYellow
 Case "Yellow":  LookUpColor = moYellow
 Case "Lime":   LookUpColor = moLimeGreen
 Case "Teal":   LookUpColor = moTeal
 Case "Dark Green": LookUpColor = moDarkGreen
 Case "Maroon":  LookUpColor = moMaroon
 Case "Purple":  LookUpColor = moPurple
 Case "Orange":  LookUpColor = moOrange
 Case "Khaki":  LookUpColor = moKhaki
 Case "Olive":  LookUpColor = moOlive
 Case "Brown":  LookUpColor = moBrown
 Case "Navy":   LookUpColor = moNavy
 End Select
End Function
```

A second function you need to add is for translating fill-style names, such as Solid Fill, into their respective integer values.

2 Enter the following in *modUtility.bas.*

```
Public Function LookUpStyle(ByVal style As String) As Integer
 Select Case style
  Case "Solid Fill":  LookUpStyle = 0
  Case "Transparent":  LookUpStyle = 1
  Case "Horizontal":  LookUpStyle = 2
  Case "Vertical":   LookUpStyle = 3
  Case "Upward Diagonal": LookUpStyle = 4
  Case "Downward Diagonal": LookUpStyle = 5
  Case "Cross":    LookUpStyle = 6
  Case "Diagonal Cross": LookUpStyle = 7
 End Select
End Function
```

With these two functions now defined, you can create the batch map. The *MakeBatchMap* sub must contain code for loading the proper map layers and rendering them, handling interactive requests such as zoom/pan and identify, and creating an output file web browsers can display. The next three sections present the code for accomplishing each of these tasks.

MakeBatchMap: Part 1

The following sub creates the map requested by the user. The values read in from the input file (for now, *test.txt*) are used to determine what colors, variables, and the like should be used. However, you must first declare some variables and load the data layers to be used.

NOTE: *You might need to change the path to the shape files to match the drive and directory where you stored the sample data.*

1 Enter the following in form *Form1*'s code page.

```
Private Sub MakeBatchMap()
 Dim stateLayer As MapObjects2.MapLayer
 Dim startColor, endColor As Long
 Dim shadestyle As Integer
 Dim uniquevals As New MapObjects2.Strings
 Dim recs As New MapObjects2.Recordset
 Dim i, j As Integer
 Dim outpath As String
 Dim outfile As String
 Dim nRecs, nMissing, nReal as Integer
 Dim stats as MapObjects2.Statistics
 Dim curval as Double
 Dim isend as Boolean
 Form1.Show
 Call Form2.addShapeFile("d:\esridata\usa", "states.shp")
   'You might need to change the path
 Call Form2.addShapeFile("d:\esridata\usa", "roads.shp")
   'You might need to change the path
 Call Form2.addShapeFile("d:\esridata\usa", "cities.shp")
   'You might need to change the path
 Map1.Layers("cities").Visible = citiesOn
 Map1.Layers("roads").Visible = roadsOn
 Map1.Layers("states").Visible = True
 Set stateLayer = Map1.Layers("States")
```

The use of the declared variables will become clear as you proceed. The first action in this sub is to place the main form, the one with the map, on the screen. Map layers are then added to the form. However, you have to make one change in how *addShapeFile* is declared in form *Form2*'s code.

2 Change this declaration from Private to Public as follows (the change is indicated in bold).

```
Public Sub addShapeFile(basepath As String, shpfile As String)
```

Returning to *MakeBatchMap* in form *Form1*, you need to set the visibility for the *cities* and *roads* layers. The layer *stateLayer* is the one for which you must set the renderer. The type of rendering is stored in the variable *mapMethod*. You also need to set the rendering options for the current *mapMethod*. The following code accomplishes this. Note that this code is similar to the code used in *frmDrawProps*, except that you are using the variables sent by the client rather than reacting to mouse click events on forms.

3 Enter the following in the *MakeBatchMap* sub.

```
Select Case mapMethod
  Case "single"
   startColor = LookUpColor(mapClr)
   shadestyle = LookUpStyle(mapStyle)
   With stateLayer.symbol
    .color = startColor
    .style = shadestyle
   End With
  Case "unique"
   shadestyle = LookUpStyle(mapStyle)
   Set recs = stateLayer.Records
   recs.MoveFirst
   Do While Not recs.EOF
    uniquevals.Add recs(mapVar).Value
    recs.MoveNext
   Loop
   Dim vr As New MapObjects2.ValueMapRenderer
   Set stateLayer.Renderer = vr
   With vr
    .ValueCount = uniquevals.Count
    .Field = mapVar
    For i = 0 To .ValueCount - 1
     .Value(i) = uniquevals(i)
     .symbol(i).SymbolType = shadestyle
    Next i
   End With
  Case "quantiles"
   startColor = LookUpColor(startClr)
   endColor = LookUpColor(endClr)
   Dim rc As New MapObjects2.ClassBreaksRenderer
   Set stateLayer.Renderer = rc
   nRecs = stateLayer.Records.Count
   With rc
    .Field = mapVar
```

```
    Set recs = stateLayer.SearchExpression("featureId > -1 order _
          by " & .Field)
    recs.MoveFirst
    Set stats = recs.CalculateStatistics(.Field)
    curval = stats.Min
    isend = False
    nMissing = 0
    .BreakCount = Val(mapClasses) - 1
    nRecs = stateLayer.Records.Count
    Do While Not recs.EOF
     If recs(.Field).Value = Null Then
      nMissing = nMissing + 1
      recs.MoveNext
     Else
      Exit Do
     End If
    Loop
    nReal = nRecs - nMissing
    j = 0
    For i = 0 To .BreakCount - 1
     Do While Not recs.EOF
      j = j + 1
      If j = (i + 1) * Int(nReal / (.BreakCount + 1)) Then
       curval = recs(.Field).Value
      End If
      recs.MoveNext
      If Not recs.EOF Then
       If j > (i + 1) * Int(nReal / (.BreakCount + 1)) Then
        If recs(.Field).Value <> curval Then
         .Break(i) = recs(.Field).Value
          curval = recs(.Field).Value
         Exit Do
        End If
       End If
        Else
        isend = True
       End If
     Loop
     If isend = True Then
      .BreakCount = i + 1
      Exit For
     End If
    Next
    .RampColors startColor, endColor
  End With
End Select
```

This code will create the map. Now you need to write the map to a file. For this, you will use the MO map control method *ExportMap*. If you look at the *ExportMap* command in the MO Help file, you will see that this command exports the map to a Windows bitmap or Enhanced Metafile. The *ExportMap* method takes three arguments: the type of file to be exported, the location at which the file should be written, and the scale factor (should the size of the map, in pixels, change).

4 Add the following line to your program.

NOTE: *You will need to set the path to the output file. The following directs the server software (Sambar43) to look for maps via a path to the author's output file.*

```
Map1.ExportMap moExportBMP, "c:\program files\sambar43\docs\bruce\_
   test.bmp", 1
```

This code will write a bitmap file of size 400 pixels by 400 pixels to *c:\programfiles\sambar43\docs*. The file will be named *test.bmp*.

You are not done with creation of the map just yet, but this would be a good time to test the program.

5 Compile your program by selecting File > Make.

6 Name the program whatever you want. (Here, it is named *mymap.exe*.)

At the beginning of this chapter, you were instructed to create a file named *test.txt*. You need to know where this file is located. Let's assume it is in *d:*. (If you placed it somewhere else, you can make the necessary changes to the following instructions.)

7 Click on the Windows Start button and then on Run.

8 Enter the name of your program (*mymap.exe*), including its path.

9 Skip a space and add the *test.txt* file, with its path. The following is an example.

```
d:\vbprograms\chapter14_1\mymap.exe
d:\test.txt
```

This should create a bitmap file, *test.bmp*, in the output directory (wherever you wrote it). This file should look like that shown in figure 14-3.

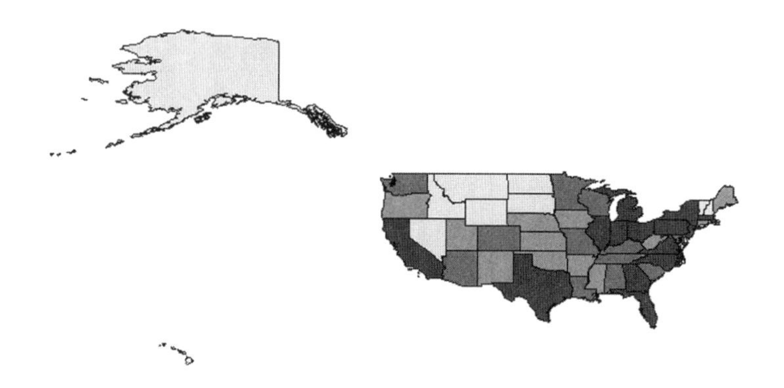

Fig. 14-3. Bitmap file created from test.txt.

You have now created a program that can read a set of input parameters stored in a file, create the corresponding map, and then export that map to a Windows bitmap file in a directory of your choosing. These abilities are necessary to respond to the requests from the client. That is, when the server receives a request from the client, it must read those requests, create the proper map, and save it to a file. There is, however, much left to be done.

MakeBatchMap: Part 2

CD-ROM NOTE: *The VB project in the* Chapter14_2 *directory on the companion CD-ROM starts here.*

This section continues the process of creating a batch map. The previous section did not implement the *Zoom*, *Pan*, or *Identify* methods. These are developed now. You begin this part of *MakeBatchMap* by declaring some new variables. This code should start immediately after the *End Select* line for the renderers and before the *Map1.ExportMap* command.

1 Enter the following.

```
Dim r As New MapObjects2.Rectangle
Dim newmapX, newmapY As Double
Dim maxmapX, maxmapY As Double
Dim bottomval, heightVal As Double
Dim zoomfactor As Single
  Set r = Map1.FullExtent
Set Map1.Extent = r
  maxmapX = Map1.Width
maxmapY = Map1.Height
 'We set this in the map1 control's property
```

```
box
newmapX = newmapY = 0
```

This code uses the rectangle, *r*, to set the map extent; that is, to handle the zoom requests. The map will be centered (assuming the user has zoomed in) at the location stored in *newmapX* and *newmapY*. *bottomVal* and *heightVal* are used in the calculation of *newmapX* and *newmapY*, as are *maxmapX* and *maxmapY*. *maxmapX* and *maxmapY* are set to the size of the map control, in pixels, and *newmapX* and *newmapY* are initialized to 0.

Once the previous code is executed, the program then needs to determine if this is the first time the user has drawn the map. If it is, no values are read in for *minX*, *minY*, *maxX*, and *maxY* (the map extent in latitude and longitude, as strings), and these values will be equal to their default values (0) set in *GetMappingValues*.

NOTE: *Here, you are taking advantage of the fact that no part of the United States is at longitude 0. If you are working in another part of the world, you would have to change the* If *condition in the following code.*

If this is *not* the first time the user has drawn the map, the program needs to set the map extent to its current map extent. The program then needs to calculate where the user clicked on the map, translating pixels to decimal degrees.

2 Enter the following immediately after the code in step 1.

```
If CInt(minX) = 0 Then
 newmapX = Map1.Extent.Center.X
 newmapY = Map1.Extent.Center.Y
Else 'translate the current map extent
 r.Bottom = CDbl(minY)
 r.Left = CDbl(minX)
 r.Top = CDbl(maxY)
 r.Right = CDbl(maxX)
 'Calculate the new center of the map
 newmapX = r.Left + ((CDbl(curX) / maxmapX) * r.Width)
 newmapY = r.Bottom + (((maxmapY - CDbl(curY)) / maxmapY) *_
   r.Height)
End If
```

Let's take a closer look at the calculation of *newmapX*. Suppose the current extent of the map, in longitude, is from –81 to –83, and that the user has clicked the mouse at X = 100, Y = 150. In this case, *r.Left = -83*. The map size in pixels is 400 x 400. Therefore, the calculation is as follows.

NewmapX = -83 + (100/400) * 2 = -82.5 degrees longitude

Notice that the calculation of *newmapY* is more complex. This is because, in the pixel world, Y increases in a downward direction.

Now that the map extent is set (in decimal degrees) and the location of the map click is converted to decimal degrees, you can execute the *Zoom/Pan* or *Identify* request. The following code executes the "zoom" functionality.

3 Enter the following immediately after the code in step 2.

```
If (mapPzi = "zoom") Then
 zoomfactor = CSng(mapLevel)
 If (CInt(zoomfactor) <> 1) Then
  Set Map1.FullExtent = stateLayer.Extent
  Set r = Map1.FullExtent
  zoomfactor = 1 / zoomfactor
  r.ScaleRectangle (zoomfactor)
  Map1.Extent = r
  Map1.CenterAt newmapX, newmapY
 Else 'zoom factor is 1
  Set Map1.FullExtent = stateLayer.Extent
  Set r = Map1.FullExtent
 End If
```

If the user zooms in or out on the map, the program converts the value of *mapLevel* to a number. If it is not equal to 1 (full extent), the program gets the full extent of the *states* layer, and sets the rectangle *r* equal to that extent. The program then takes the inverse of *zoomfactor* and scales the rectangle by that amount. In MO, when you scale the rectangle by a number less than 1, you essentially tighten the zoom window. The program sets the map extent to the rectangle extent, and then centers the map where the user clicked (converted to latitude/longitude).

If the zoom factor is 1, the program simply gets the full extent of the *states* layer and sets the map extent equal to that. The other option to consider is whether the user has chosen the Identify option. If so, the program needs to find the state on which the user clicked.

4 Enter the following.

```
Else 'mapPzi = "Identify"
  Map1.Extent = r
```

```
  Dim webSelection As MapObjects2.Recordset
  Dim tol As Double
  Dim pt As New MapObjects2.Point
  tol = Map1.ToMapDistance(5) 'You can use a different number
  pt.X = newmapX
  pt.Y = newmapY
  Set webSelection = stateLayer.SearchByDistance(pt, tol, "")
  If webSelection.Count > 0 Then
   Set gSelection = webSelection
  Else
   Set gSelection = stateLayer.SearchExpression("featureID = -1")
  End If
End If
```

This section is similar to the code you used for the Identify button, and it should be! The map is set to its current extent (read in from the web page). After declaring some necessary variables, the program sets the search tolerance to 5 pixels. (You can use a different tolerance value if you wish.) The program then creates a point whose coordinates correspond to where the user clicked on the map. If any records are chosen, the program sets *gSelection* to the selected set, as was done for the Identify button. Next, the map is refreshed, which forces a call to *AfterLayerDraw* so that MO can highlight the selected state.

In order to support batch mapping, you need to change one line in *AfterLayerDraw*. The following code shows this.

5 Change

```
If Toolbar1.Buttons("Identify").Value = tbrPressed Then
```

to

```
If (Toolbar1.Buttons("Identify").Value = tbrPressed) Or (mapPzi = _
   "identify") Then
```

6 Return to the *MakeBatchMap* sub and add the following call to *Map1.Refresh* before exporting the map. (You will need to put in the correct path for the output map.)

```
Map1.Refresh
Map1.ExportMap moExportBMP, "c:\program files\sambar43docs\bruce\_
   test.bmp", 1
```

This completes all drawing and map interaction functions you need to create the requested map. Let's test these new functions before developing the code for writing the resultant web page.

7 Change the *test.txt* file to read as follows.

```
00000
variable
pop1990
method
unique
style
Solid Fill
level
5
X
310
Y
110
minx
-91.21
miny
31.53
maxx
-83.28
maxy
39.39
pzi
identify
variable
Sub_region
```

If you run the program again, using this version of *test.txt*, your map should look like that shown in figure 14-4.

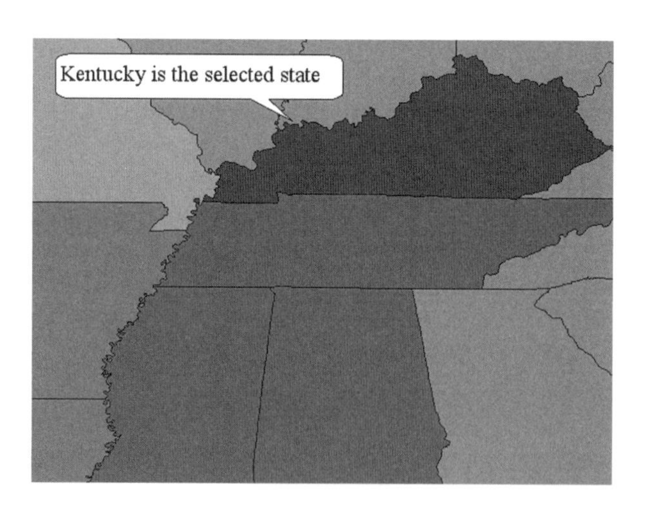

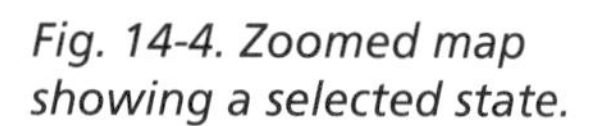
Fig. 14-4. Zoomed map showing a selected state.

MakeBatchMap: Part 3

In this section you will use VB to create an image file that can be displayed in a web browser. The first step in building the web page is to create a *jpg* file. Web browsers using only HTML cannot show bitmap files, but they can display *gif* and *jpg* files.

1 On the companion CD-ROM, in the *utilities* directory, is a DLL named *jpgegdll.dll*. Copy this to your hard disk.

In the code that follows, it assumed that this DLL resides in a directory named *c:\gistools\dll*. This DLL is based on code from the Independent JPEG Group. This source code is free, and can alternatively be downloaded from the Web.

As you saw in Chapter 2, DLLs expose functions to other programs. This DLL contains a function called *Convert*. It takes two arguments: the name of the input *bmp* file, and the name of the output *jpg* file. These names should include the full path to the files.

To use an external DLL in a VB program that is not a plug-in (such as that used in Chapter 2), you have to include the proper declaration.

2 For the *jpgegdll.dll*, enter the following declaration in the *modUtility.bas* code page, in the "declarations" section.

```
Public Declare Function Convert Lib "c:\gistools\dll\jpegdll.dll"_
   (ByVal inputbmp As String, yVal outputjpg As String) As Long
```

You will need to change the path if you have placed the DLL elsewhere. You can now convert the bitmap created by the program to a *jpg* file.

3 Return to the *MakeBatchMap* sub and delete the current call to the *ExportMap* line and replace it with the following lines of code.

NOTE: *Again, this assumes you are using Sambar Web Server. If you are not, you may need to change the path in* theBmp *and* theJpg.

```
Dim theBmp As String
Dim theJpg As String
theBmp = "c:\Program Files\sambar43\docs\bruce\" & jobID & ".bmp"
theJpg = "c:\Program Files\sambar43\docs\bruce\" & jobID & ".jpg"
Map1.ExportMap moExportBMP, theBmp, 1
Convert(theBmp, theJpg)
```

Your program can now create a properly rendered map at a requested scale in a format that can be displayed by a web browser. You now need to develop a web page to display the resulting map. This process will implement many of the *html* functions discussed in Chapter 13.

Writing the Web Page

In this section, you will write the web page. There are several things you need to know. You need to know your web server address. You also need to know where you want to store web pages and images on the server, and where your VB/MO executable file is located, along with the bitmap to the *jpg* DLL. This chapter uses the web address *http://www.yourwebaddress.edu.*

You need to substitute your web address wherever you see this address. In the following code, you will create an *html* form that has the map as an action button. The *action=* section for the form refers to a PERL script, developed in the following. In addition, paths to the VB/MO program and the DLL are used that may not match those on your server. You may have to update those, too. Finally, you will need to change the location of where the web pages, bitmaps, and *jpg*s are written. Because *MakeBatchMap* is getting a bit long, you will create a new sub for writing the web page.

1. Place the line indicated in bold in the following as the last line in *MakeBatchMap.*

```
 Map1.ExportMap moExportBMP, theBmp, 1
 outValue = Convert(theBmp, theJpg)
 MakeWebPage
End Sub
```

Now you can create the web page. In the new sub, *MakeWebPage,* you will need to declare two strings.

```
Private Sub MakeWebPage()
 Dim outpath As String
 Dim outfile As String
 Dim i As Integer
 outpath = "c:\Program Files\sambar43\docs\bruce\"
 'You may need to change this
 outfile = outpath & jobID & ".html"
 Open outfile For Output As #2
```

This sets the path to where the web page should be written. If the directory where you store server pages is different than that specified in this code, you will need to change *outpath.*

You can now use the Print directives in VB to write the web page. The new page will include a title and a form, and the form action will call a PERL script that creates the map (the script is found on the companion CD-ROM).

2 Enter the following.

```
Print #2, "<HTML><HEAD>"
Print #2, "<TITLE>Developing GIS Web Pages, Strategy 1</TITLE>"
Print #2, "<FORM ACTION=""http://www.yourwebaddress.edu/cgi-bin/_
  method1.pl"" METHOD = POST>"
Print #2, "</HEAD><BODY>"
```

The body of the page will contain a set of hidden inputs. These do not have to be at the start of the page, but they do need to be somewhere in the body. The hidden inputs will depend, in part, on the type of map constructed.

3 Enter the following.

```
Print #2, "<INPUT TYPE=HIDDEN NAME=""method"" VALUE=""" & mapMethod & """>"
Select Case mapMethod
  Case "single"
   Print #2, "<INPUT TYPE=HIDDEN NAME=""color"" VALUE=""" & mapClr & """>"
   Print #2, "<INPUT TYPE=HIDDEN NAME=""style"" VALUE=""" & mapStyle & """>"
   Print #2, "<H1>Single Value Map</H1>"
  Case "unique"
   Print #2, "<INPUT TYPE=HIDDEN NAME=""style"" VALUE=""" & mapStyle & """>"
   Print #2, "<INPUT TYPE=HIDDEN NAME=""variable"" VALUE=""" & mapVar & """>"
   Print #2, "<H1>Unique Value Map of " & mapVar & "</H1>"
  Case "quantiles"
   Print #2, "<INPUT TYPE=HIDDEN NAME=""variable"" VALUE=""" & mapVar & """>"
   Print #2, "<INPUT TYPE=HIDDEN NAME=""classes"" VALUE=""" & mapClasses & """>"
   Print #2, "<INPUT TYPE=HIDDEN NAME=""start"" VALUE=""" & startClr & """>"
   Print #2, "<INPUT TYPE=HIDDEN NAME=""end"" VALUE=""" & endClr & """>"
   Print #2, "<H1>Quantile Map of " & mapVar & "</H1>"
End Select
```

This code prints out the map method and the variables specific to it as hidden inputs. For example, in the unique-value map, you pass *mapStyle* and *mapVar* as hidden. The first thing the user sees on the page is a headline indicating the type of map requested. For example, the first line on the page shown in figure 14-1 was generated by the following line, with *mapVar* equal to *sub_Region.*

```
Print #2, "<H1>Unique Value Map of " & mapVar & "</H1>"
```

The next section of the web page gives the user the choice of using the Identify or the Zoom/Pan option, which is the default. It also lets the user set the zoom level.

4 Enter the following.

```
Print #2, "The next mouse click will: "
Print #2, "<INPUT TYPE=RADIO NAME=""pzi"" VALUE=""Identify""> Identify "
Print #2, "<INPUT SELECTED TYPE=RADIO NAME=""pzi"" VALUE=""Zoom"" CHECKED>_
    Zoom And Pan "
Print #2, "<SELECT NAME=""level"">"
For i = 1 To 10
  If i = CInt(mapLevel) Then
    Print #2, "<OPTION SELECTED VALUE=""" & i & """>" & i & "</OPTION>"
  Else
    Print #2, "<OPTION VALUE=""" & i & """>" & i & "</OPTION>"
  End If
Next
Print #2, "</SELECT>"
Print #2, "<BR>"
```

The *For* loop fills the selection box for zoom level, making sure that the selected value is equal to the current *mapLevel*. The next section of code sends the current map extent as hidden inputs. You need these so that the program can calculate the value of the *x,y* coordinates of the user's next mouse click on the map. (See previous discussion.) The program also needs to send the values of *roadsOn* and *citiesOn* as hidden inputs.

5 Enter the following.

```
'send the current map extent
Print #2, "<INPUT TYPE=HIDDEN NAME=""minx"" VALUE=""" & Map1.Extent.Left _
      & """>"
Print #2, "<INPUT TYPE=HIDDEN NAME=""miny"" VALUE=""" & Map1.Extent.Bottom_
      & """>"
Print #2, "<INPUT TYPE=HIDDEN NAME=""maxx"" VALUE=""" & Map1.Extent.Right _
      & """>"
Print #2, "<INPUT TYPE=HIDDEN NAME=""maxy"" VALUE=""" & Map1.Extent.Top _
      & """>"
If roadsOn Then
  Print #2, "<INPUT TYPE=HIDDEN NAME=""roads"" VALUE=on>"
End If
If citiesOn Then
  Print #2, "<INPUT TYPE=HIDDEN NAME=""cities"" VALUE=on>"
End If
```

Now you can serve the map. To make sure the map has a border around it, you will embed it in a bordered table. The table will consist of one cell—the map. You can access the map using a relative path. That is, the path to the map is relative to where the server software (such as Sambar, Apache, or Microsoft's Personal Web Server or IIS) is located.

6 Enter the following.

```
 Print #2, "<TABLE BORDER>"
 Print #2, "<TR><TD>"
 Print #2, "<INPUT TYPE=IMAGE SRC=/docs/bruce/" & jobID & ".jpg > "
 Print #2, "</TR></TD></TABLE>"
 Print #2, "<BR>"
```

If the map is being generated in response to an *Identify* request, the program needs to list the information for the selected states in a table. A hyperlink is placed in this table so that the user can go to the selected state's official web page. The *states* layer contains a field called *state_abbr* that contains the two-letter abbreviation of each state. Each state's official web page has a URL of the form *ww.state.abbr.us,* where *abbr* is the two-letter abbreviation of an individual state.

7 Enter the following code for writing out the selected records.

```
If mapPzi = "identify" Then
 If gSelection.Count = 0 Then
  Print #2, "No states were selected"
 Else 'we have at least one selected set.
  Dim tabDesc As New MapObjects2.TableDesc
  Dim curfield As New MapObjects2.Field
  gSelection.MoveFirst
  Do While Not gSelection.EOF
   Set tabDesc = gSelection.TableDesc
   Print #2, "<TABLE BORDER>"
   For j = 0 to tabDesc.FieldCount - 1
    Set curfield = gSelection.Fields(tabDesc.fieldName(j))
    Print #2, "<TR>"
    Print #2, "<TD> " & curfield.Name & "</TD>"
    If curfield.Type = moString Then
     If UCase(curfield.Name) = "STATE_ABBR" Then
   Print #2, "<TD><A HREF=""http://www.state." & curfield.Value & _
          ".us"">" & curfield.Value & "</a>"
     Else
      Print #2, "<TD> " & curfield.Value & "</TD>"
     End If
    Else
     Print #2, "<TD> " & curfield.ValueAsString & "</TD>"
    End If
   Next j
```

```
    Print #2, "</TABLE><BR>"
    gSelection.MoveNext
   Loop
  End If
End If
```

This code begins by verifying that the currently selected action (*mapPzi*) is *identify*. It then checks to see if there are any selected records. If not, it prints a message stating this fact to the user. If there are selected records, the program goes to the first record and starts building tables.

The *Do While* loop starts by creating the selected state's table. The table will be very similar to the *fieldList* box of the *SelectedForm* you developed in Chapter 7. That is, each row will report two items: a database field name and its value. You will place each of these in a separate column.

There is one difference between this table and the items reported for the Identify button. In this instance, the program will report only the standard database fields, not the *featureID* or the Shape field. For this, you use the *tableDesc* object, which corresponds to the selected record set. The program cycles through each item in the *tableDesc* (for *j = 0* to *tabDesc.FieldCount – 1*) and sets the current field (a *MapObjects Field* object) to the field with the same name as the *j*th field in *tableDesc.*

For each selected state, the program declares a bordered table. The program then cycles through each field in the selected set, adding a row, and then the column for the field name. It then adds a column for the field's value. However, you have to be careful. First, the program needs to check if the field is a string or not. If it is, its value is written. If it is not, its value is written as a string.

In the first case, a string field, the program also needs to check to see if it is the state abbreviation. If that is the case, the field's value is listed and a hyperlink is associated with that state's official web page. The program continues through all fields until it has reported all of them. The program then gets the next selected state, if there is one. The last operations you need to perform are to finish the web page, close the file, and end the sub.

8 Enter the following.

```
 Print #2, "</FORM></BODY></HTML>"
 Close #2
End Sub
```

You now have the ability to create web pages that look like that shown in figure 14-5.

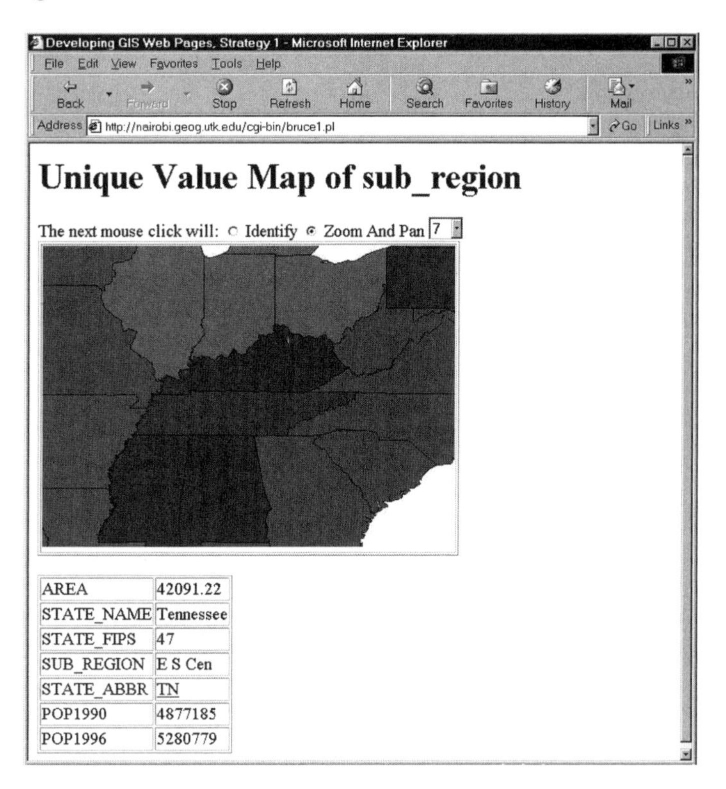

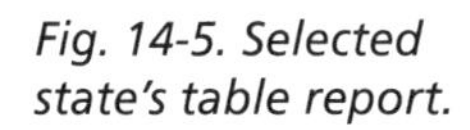
Fig. 14-5. Selected state's table report.

Putting It All Together

CD-ROM NOTE: *The VB project in the* Chapter14_3 *directory on the companion CD-ROM starts here.*

The web site you will develop will use a setup form to determine the initial parameters for mapping, as shown in figure 14-6.

A copy of this page, *form1.html*, is found on the companion CD-ROM in the *web* directory. The page consists of three hyperlinks to anchors on the page. The anchors correspond to the type of map renderer the user wants to use: a single-value map, a unique-value map, or a quantile map.

Each anchor sets off a form. Each form will call the same PERL script. The script that is called should be in the *cgi-bin* directory on your server. Therefore, you need to place the correct address in

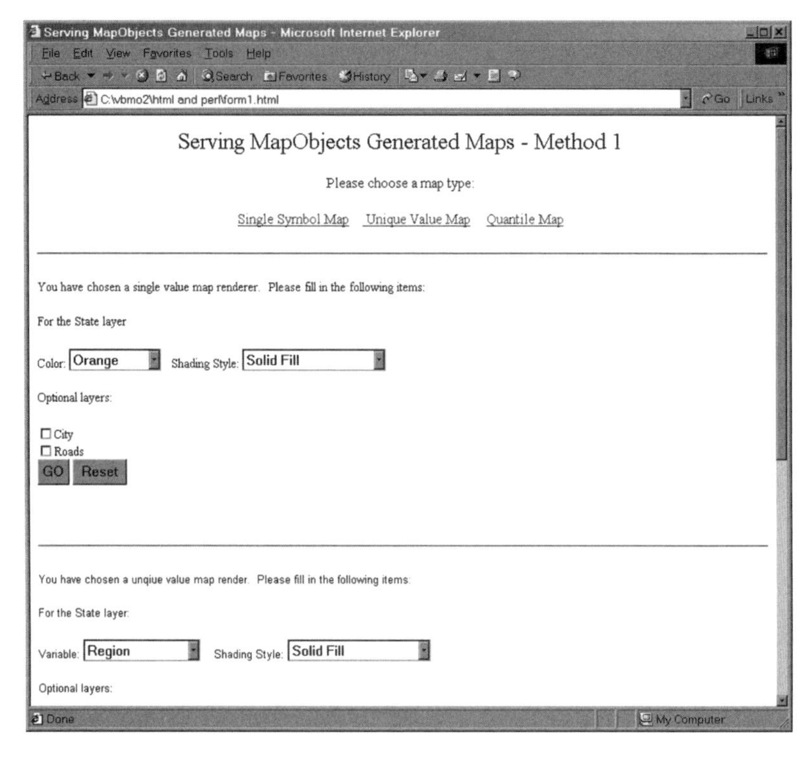

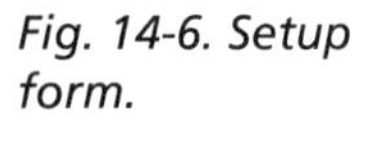
Fig. 14-6. Setup form.

the *html* code. That is, you need to change the following address to your server address.

1 Change all instances of the following address to reflect your web address.

```
<form ACTION="http://www.yourwebaddress.edu/cgi-bin/method1.pl" _
   METHOD="POST">
```

For example, if your web address is *www.mymaps.com* and your PERL scripts are kept in the *cgi-bin* directory, this line (in all three instances) should read as follows.

```
<form ACTION="http://www.mymaps.com/cgi-bin/method1.pl" _
   METHOD="POST">
```

When the user chooses to create a map, the inputs from the web page are sent to the server and passed to a PERL script named *method1.pl*. That script follows. The following does not attempt to explicate every line of this code. If you are interested in learning more PERL, visit *http://www.perl.com/CPAN-local/doc/manual/html/pod/perlfunc.html*, the ActiveState site mentioned at the beginning of this chapter. In addition, consult *PERL 5 Complete* by Peshko and DeWolfe. The following is the PERL script, followed by a brief description of what the script does.

```
read(STDIN,$temp,$ENV{'CONTENT_LENGTH'});
$| = 1;
print "Content-type: text/html\n\n";
print "<pre>";
@pairs=split(/&/,$temp);
foreach $item(@pairs) {
 ($key,$content)=split (/=/,$item,2);
 $content =~ tr/+/ /;
 $content =~ s/%(..)/pack("c",hex($1))/ge;
 $fields{$key}=$content;
}
print "</pre>";
srand;
$uniq = rand;
$uniq1 = $uniq * 100000;
$uniq2 = int($uniq1) ;
$jobid = $uniq2 ;
$| = 1;
open($jobid,">c:\\progra~1\\sambar43\\docs\\bruce\\$jobid.txt");
print $jobid "$jobid\n";
foreach $varkey ( keys %fields) {
 print $jobid "$varkey\n" ;
 print $jobid "$fields{$varkey}\n" ;
}
close ($jobid);
system ("c:\\brucevb\\mymap.exe c:\\progra~1\\sambar43\\docs\\_
   bruce\\$jobid.txt");
sleep 2;
open(OUTPAGE,"c:\\progra~1\\sambar43\\docs\\bruce\\$jobid.html");
while(<OUTPAGE>) {
      $line = $_;
      print $line;
}
```

The script generates a random number to use as a process ID (the variable name for this is *$jobid*) for the current request. It then reads the *standard in* input stream (that is, what the web page returns) and splits that line of information into pairs: the variable name (*$varkey*) and its value (*$fields($varkey)*). These values are written to a text file with the name *$jobid.txt*. Therefore, if the value of *$jobid* is 00000, the file written will be *00000.txt*.

Once the file is written, the PERL script calls the VB/MO program with the file name (*$jobid.txt*) as a command line argument. It then waits for 2 seconds (*sleep 2*) while the new map and web page are created. It then reawakens, reads the web page created, and line-by-line pushes it back to the client.

There are three things you will need to change in this script. The first is where to write the *$jobid.txt* file. In the script, this is implemented as follows.

```
open($jobid,">c::\\progra~1\\sambar43\\docs\\bruce\\$jobid.txt");
```

2 Change the portion of the previous code so that the path to the *$jobid.txt* file points to where you will store this file on your server.

NOTE: *Leave the double slashes in. PERL, like C, interprets a slash as a maker of special characters. The double slash tells PERL that the special character you want is a slash.*

The next line to change is the call to the VB/MO program, along with the path to the input file (a command line argument). The script currently has the following values.

```
system ("c:\\brucevb\\mymap.exe c:\\progra~1\\sambar43\\docs\\ _
   bruce\\$jobid.txt");
```

The first thing you must check is that the path to the VB/MO program is correct. The second thing you need to check is that the path to the file *$jobid.txt* matches the one you used to write that file. Suppose, for example, your VB/MO program were in *c:\myvb* and had the name *supermap.exe*. You would replace

```
c:\\brucevb\\mymap.exe
```

with

```
c:\\myvb\\supermap.exe
```

Now suppose that the *$jobid.txt* file were written to *c:\webshare\wwwroot\temp*. You would replace

```
c:\\progra~1\\sambar43\\docs\\bruce\\$jobid.txt
```

with

```
c:\\webshare\\wwwroot\\temp\\$jobid.txt
```

The result would be a call that reads

```
system ("c:\\myb\\supermap.exe c:\\webshare\\ wwwroot\\temp\\$jobid.txt");
```

The final thing you need to change in the PERL script is the line that opens the HTML file your VB/MO program generated (the one that will be pushed back to the client). Currently, this line is

```
open(OUTPAGE,"c:\\progra~1\\sambar43\\docs\\bruce\\$jobid.html");
```

You will need to change the path to this file so that it matches the path used in your VB/MO program. That path was specified in the VB/MO program with the line

```
outpath = "c:\Program Files\sambar43\docs\bruce\"
```

Summary

This chapter presented a method of serving maps on the Web that used a small PERL script for passing values from the client to your VB/MO program. The program you created can run as an interactive program or as a batch program. You tested the page as you went along, using a file, *test.txt*, to simulate returns from a client. You implemented zoom/pan and identify options and wrote them to a web page. Finally, you used a setup form and a PERL script to put all of the parts together. As you build your programs, you will need to change the address of the web page to match your server, and the paths to file locations to match your particular directory structure.

There is nothing unique about using PERL. There are other programming languages you could have used to write a program to do the same job as this PERL script. There are numerous books, such as *CGI Programming with C and C++* by Felton, on this topic. PERL does have the advantage that it can be edited in any text editor.

The current method will write several files to your server's hard disk. You will need to clean these out periodically. If you have a web administrator, she can probably write a small utility program to do this.

In the next chapter you will work with the MapObjects Internet Map Server (MOIMS). It handles things a bit differently, but there are many similarities between how the MOIMS works and how you served maps in this chapter. In fact, you will only have to make a few changes to the VB/MO program to switch from this method to the MOIMS method.

Chapter 15

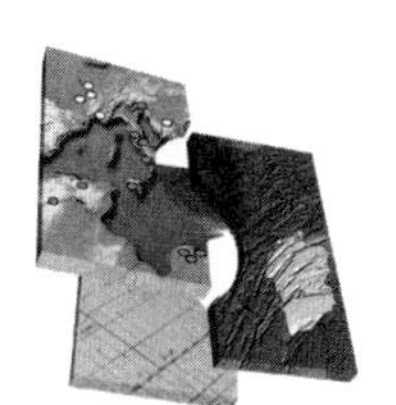

Serving Maps on the Web: Method 2

ESRI's Internet Map Servers

■■ Introduction

In this chapter you will study other methods of serving maps on the Web. These methods involve using ESRI's Internet Map Servers (IMS). The MapObjects (MO) IMS (MOIMS) and ArcIMS are both discussed in this chapter.

You will recall from the previous chapter that the project you developed worked as follows. The user fills out a form (*form1.html*) and submits it. A PERL script reads the input and generates a file. The PERL script then issues a system call that initiates the VB/MO program. That program reads the file of user inputs, creates the proper map, converts it to a *jpg* file, writes the new web page, and then shuts down. The PERL script grabs the web page generated by the program (which includes the map) and pushes that back to the user.

NOTE: *The VB/MO program does not have to be on the same computer as the server software. However, the server does need to be able to issue the proper call to the program and read its output.*

This approach works, but it is not the only way. Another way is to use an IMS. The following section takes a brief look at the MO IMS, including its major components and an example of how to use it.

NOTE: *For a more in-depth discussion of the IMS, see ESRI's* MapObjects IMS User's Guide. *You can download this file at* http://www.esri.com/software/mapobjects/ims/moims_

download.html. *Just click on* MapObjects IMS User's Guide *to get the file in* pdf *format.*

An Overview of the MOIMS

The MOIMS consists of several parts. You will work most closely with two of these: a DLL (*esrimap.dll*) and an ActiveX control. The DLL is used to handle requests from clients, parse them, and pass them to servers running VB/MO programs. It also accepts responses from VB/MO programs (maps and web pages) and passes them back to the client.

The ActiveX control, called a WebLink control, gets imbedded into a VB project like any other control. This control manages the conversation between the VB/MO program and the DLL *esrimap.dll*. That is, it receives messages (map requests) from the DLL and sends messages (the *html* page) back to the DLL, which then passes them on to the client. It does a few other things, too, such as converting bitmaps to *jpg* or *gif* files.

The MOIMS architecture uses a three-tiered approach. The first tier is the client. It is here the user fills out a form or interacts some other way with a web page. Once the user submits that information, it is passed to the middleware (MOIMS). The DLL *esrimap.dll* serves a function similar to the PERL script of Chapter 14. It takes the requests from the client, parses them, and then passes them to the proper application.

In the MOIMS world, all programs to receive requests are always running on the server, and the client sends messages to the server using the *Get* method rather than the *Post* method. Once the programs generate the proper map, it is pushed back to the client, along with the *html* for the web page. The middleware manages these operations.

The MOIMS contains some special communication abilities. In particular, it can communicate with MO programs (such as the one you have been working with) that contain a WebLink control. The MOIMS can also communicate with a server running an ArcExplorer project. As you may know, ArcExplorer is freeware that allows you to use projects (similar to ArcView projects) on the Web. On the server side, you can set which layers, if any, the client can download from the server (i.e., you can set up a spatial data *ftp* site).

The third tier in this approach is the server tier. This is where programs, such as ArcExplorer or your VB/MO programs, run. As mentioned previously, all programs the IMS interacts with must be running constantly.

There is an advantage to this approach. The method used in Chapter 14 required the program to be loaded into memory and unloaded from memory for every request. This involves quite a bit of overhead and can slow processing time. The MOIMS approach should run faster. To use the MOIMS, you must integrate the WebLink control in the VB/MO program and implement its communication and data-handling features.

Working with the WebLink Control

CD-ROM NOTE: *The VB project in the* Chapter15_1 *directory on the companion CD-ROM starts here.*

In this section you will work on a copy of the project you developed in Chapter 14. With minor changes, much of the code used in that chapter is reusable here.

1. Add a control (Ctrl-T) to your project: the ESRI MapObjects WebLink control (figure 15-1).
2. Add a WebLink control to the form (*Form1*) that contains the map control.

Like some other controls you have used, this is a hidden control (figure 15-2).

This control has several properties, the most important of which is the *MapPort* property. This property represents the port the WebLink control uses to communicate with the DLL *esrimap.dll.*

The WebLink control has several methods, among which are *Start, Stop, WriteResponseHeader, WriteString, BMP2JPEG,* and *BMP2GIF.* (The control's only event is *Request.*) The following describes these methods.

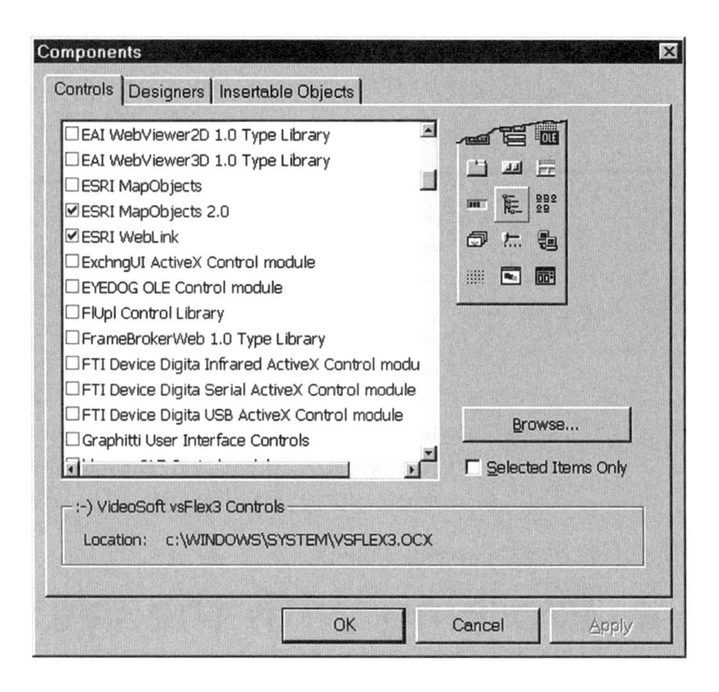

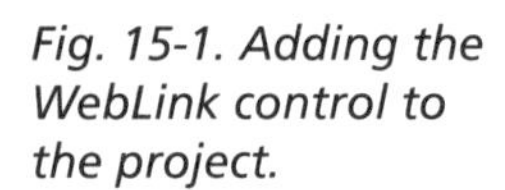
Fig. 15-1. Adding the WebLink control to the project.

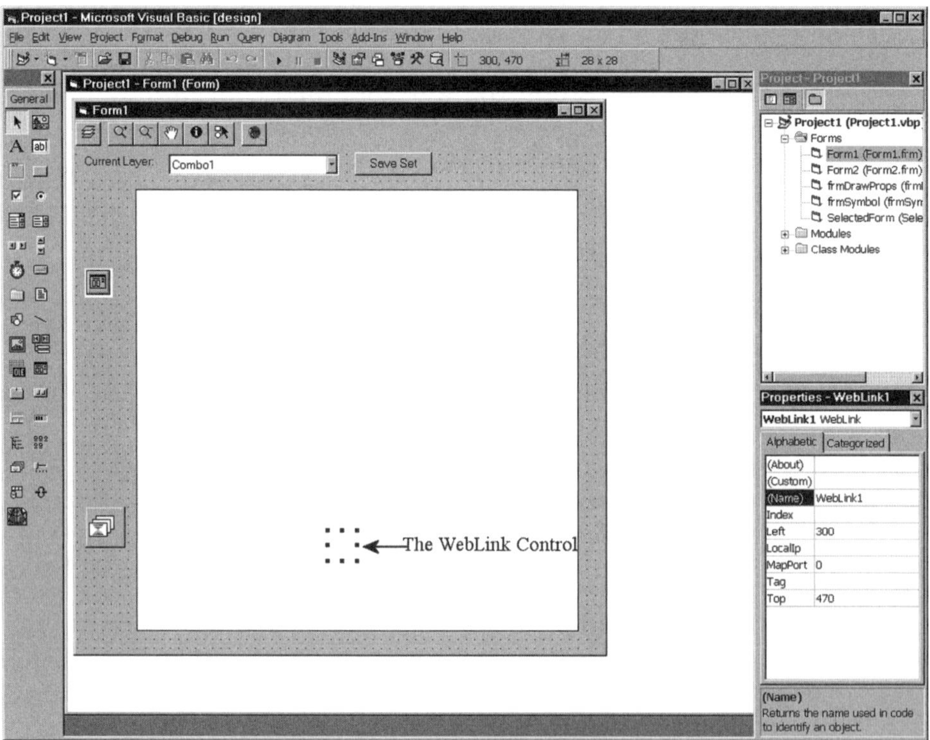

Fig. 15-2. Placing the WebLink control on form Form1.

Start is a method that tells the WebLink control to establish a conversation with the DLL *esrimap.dll*. It returns a Boolean value (True or False) that indicates whether the connection between the VB/MO program and *esrimap.dll* was established. The WebLink control "listens" via the *MapPort* property. (You can think of a map port as a radio frequency or TCP/IP address.) *Stop* ends the conversation between the VB/MO program and *esrimap.dll*.

WriteResponseHeader passes the header line to the DLL that will be passed to the client. The header line indicates that what is to follow is an *html* page. In Chapter 14, you did this in PERL using the following line.

```
print "Content-type: text/html\n\n";
```

In this chapter, you will place this line in your VB/MO program as a *WebLink* request, as follows.

```
WebLink1.WriteResponseHeader "Content-type: text/html" + vbCrLf + vbCrLf
```

In Chapter 14, you wrote out the lines of your new web page from the VB/MO program with statements such as the following.

```
Print #2, "<INPUT TYPE=HIDDEN NAME=""method"" VALUE=""" & mapMethod & """>"
```

When you employ the MOIMS here, you will replace *Print #2* in this code with *WebLink1.WriteString*, as in the following.

```
WebLink1.WriteString "<INPUT TYPE=HIDDEN NAME=""method"" VALUE=""" & _
   mapMethod & """>"
```

Instead of writing to a file, this writes directly to the DLL, which pushes it back to the client. *BMP2GIF* and *BMP2JPEG* do exactly what you would expect: they convert bitmap files to *gif* or *jpg* files. Let's get started with a project.

1. Add the following variable to the start of form *Form1*'s code page (i.e., give it module-level scope).

   ```
   Private filecounter As Long
   ```

You will use this variable to count requests for maps (more on this in material to follow).

2. Change the *Form_Load* function to read (you may need to alter the paths to match your computer) as follows.

```
Private Sub Form_Load()
  batchOn = False
  Set gSelection = Nothing
```

```
  WebLink1.MapPort = "5063"
  If WebLink1.Start Then
    batchOn = True
    Call Form2.addShapeFile("d:\esridata\usa", "states.shp")
    Call Form2.addShapeFile("d:\esridata\usa", "roads.shp")
    Call Form2.addShapeFile("d:\esridata\usa", "cities.shp")
  Else
    RefreshCombo1
  End If
End Sub
```

As in Chapter 14, this code begins by setting *batchOn* to False and *gSelection* to nothing. *MapPort* is then set. This is the port number on which the control communicates with the DLL. The program then issues a *Start* request. If the program is being used with the IMS, this should return True. If not, the program runs as an interactive program, as in previous chapters.

If serving maps with the MOIMS (i.e., *WebLink1.Start* is True), *batchOn* is set to True and all data is loaded. Can you see why you changed the location data loading? When you used PERL, you loaded the data in the *MakeBatchMap* sub. However, because in the MOIMS world the program will run constantly, you want to add the map layers just one time. If you had left the calls to *addshape* where they were in Chapter 14, every map request would load the layers. The first time someone visited the web site, there would be three map layers; the second time, six layers; and so on.

After a while the program would crash. By moving the *addshape* request to the *Form_Load* sub, these layers get added just once. When the program ends, the *WebLink* function needs to stop.

3 Create a *Form_Unload* sub in order to stop the *WebLink* function.

```
Private Sub Form_Unload(Cancel As Integer)
    If batchOn Then
        WebLink1.Stop
    End If
End Sub
```

The key function for responding to a client's request is *WebLink.Request()*. Consider this function; it will determine what happens when a request arrives from the DLL *esrimap.dll*. Functionally, this is similar to the *GetMappingValues* of Chapter 14. Thus, this sub should look similar in design to the *GetMappingValues* sub used in the previous chapter.

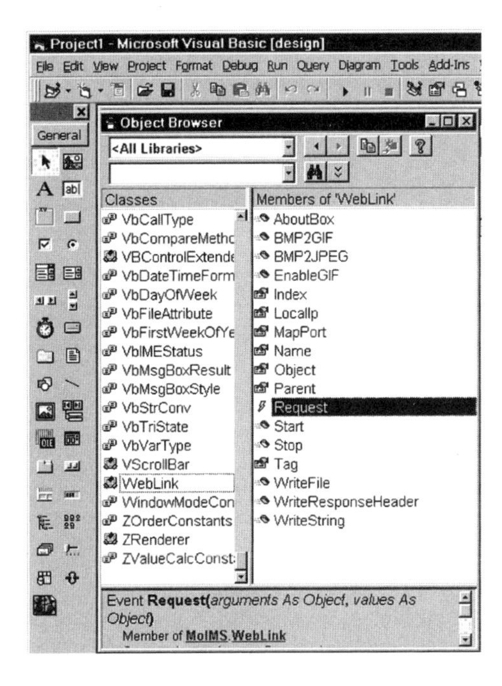

4 Start the Object browser (either click on the F2 button or on View-Object Browser), scroll to the *WebLink* object, and highlight the *Request* event. Your screen should look like that shown in figure 15-3.

The request receives two objects: arguments and values. Both of these are string collections. Each argument will have a matching value. This is very similar to what the *$jobid.txt* files contained in Chapter 14. That is, in the previous chapter, you created files that had the following structure.

Fig. 15-3. The WebLink.Request *event in the Object browser.*

```
Jobid
Variable 1
Value 1
Variable 2
Value 2
   .
   .
   .
Variable n
Value n
```

The *esrimap.dll* does not send a *jobid* number (you will generate this in material to follow). However, it does send two parallel lists that have the following structure.

```
Variable 1     Value 1
Variable 2     Value 2
Variable 3     Value 3
   .                            .
   .                            .
   .                            .
Variable n     Value n
```

Here, the first column is the argument list and the second column is the value list. This is similar to the input file you used in Chapter 14, and it should be no surprise that the *Request* function, which follows, is similar to the *GetMappingValues* function.

5 Enter the following.

```
Private Sub WebLink_Request(ByVal arguments As Object, ByVal values _
    As Object)
  Dim i As Integer
  mapLevel = "1"
  curX = "0"
  curY = "0"
  minX = "0"
  minY = "0"
  maxX = "0"
  maxY = "0"
  mapPzi = "zoom"
  jobID = "00000"
  mapLevel = "1"
  roadsOn = False
  citiesOn = False
  For i = 0 To arguments.Count - 1
     Select Case arguments(i)
     Case "pzi"
        mapPzi = values(i)
     Case "level"
        mapLevel = values(i)
     Case "minx"
        minX = values(i)
     Case "miny"
        minY = values(i)
     Case "maxx"
        maxX = values(i)
     Case "maxy"
        maxY = values(i)
     Case "roads"
        roadsOn = True
     Case "cities"
        citiesOn = True
     Case "method"
        mapMethod = values(i)
     Case "x"
        curX = values(i)
     Case "y"
        curY = values(i)
    'quantile variables
     Case "classes"
        mapClasses = values(i)
```

```
    Case "start"
       startClr = values(i)
    Case "end"
       endClr = values(i)
    'single value map
    Case "color"
       mapClr = values(i)
    Case "style"
       mapStyle = values(i)
    'quantiles or unique value
    Case "variable"
       mapVar = values(i)
    End Select
  Next
  MakeBatchMap
End Sub
```

After the variable declaration, the first part of this sub sets the default mapping values. It then cycles through every variable passed from the client; that is, through every element in the argument collection. This code uses the same Select Case approach used in *GetMappingValues*, except that this time the mapping variable is set equal to the *i*th argument (except for the Boolean values). The sub ends with a call to *MakeBatchMap*.

There are very few changes that need to be made to *MakeBatchMap*. The first is to comment out (or remove) the lines that add layers to the map.

6 Comment out all calls to *Form2.addShapeFile* in *MakeBatchMap*.

```
  ' The following lines are commented out
  ' Call Form2.addShapeFile("d:\esridata\usa", "states.shp")
  ' Call Form2.addShapeFile("d:\esridata\usa", "roads.shp")
  ' Call Form2.addShapeFile("d:\esridata\usa", "cities.shp")
```

The rest of the sub remains the same until the end. The following is the end of the sub. Lines that have changed from the previous chapter's program are in bold.

7 Make the following changes indicated in bold.

```
  Dim theBmp As String
  Dim theJpg As String
  Dim timeStamp As String
  timeStamp = Format(Now, "h_m_s") & filecounter
  theBmp = "c:\webshare\wwwroot\temp\" & timeStamp & ".bmp"
```

```
  theJpg = "/temp/" & timeStamp & ".jpg"
  Map1.ExportMap moExportBMP, theBmp, 1
  WebLink1.BMP2JPEG theBmp
  Kill theBmp
  filecounter = filecounter + 1
  MakeWebPage (theJpg)
End Sub
```

Because *esrimap.dll* does not pass in a *jobid*, as the PERL script did, you need to create a unique name for the map file this program creates. In the previous code, a time stamp and file counter are used. The *filecounter* variable is used in case more than one request gets processed at the same hour, minute, and second. After creating a new *bmp* file, the program exports this file and calls the WebLink control's *BMP2JPG* method. This creates a new *jpg* file. The call to *Kill* removes the *bmp* file (no longer needed) from the server's disk. (See the section on server cleanup in the following material.) The program then increments *filecounter* and calls *MakeWebPage.* This sub has changed, in that it now passes in the name of the *jpg* file.

Like the *MakeBatchMap* sub, the *MakeWebPage* sub is very similar to the one used in Chapter 14. It removes all references to *outfile* and *outpath.* It also writes out the *ResponseHeader,* and then replaces every instance of *Print #2* with a call to the WebLink control's *WriteString* method. The following is the new version of the sub. Changes are indicated in bold.

8 Make the following changes indicated in bold.

```
Private Sub MakeWebPage(theImage As String)
  With WebLink1
    .WriteResponseHeader "Content-type: text/html" + vbCrLf + vbCrLf
    .WriteString "<HTML><HEAD>"
    .WriteString "<TITLE>Developing GIS Web Pages, Strategy 2</TITLE>"
    .WriteString "<FORM ACTION=/scripts/esrimap.dll>" & vbCrLf
    .WriteString "<INPUT TYPE=hidden NAME=name Value=Mymap2>" & vbCrLf
    .WriteString "</HEAD><BODY>"
    .WriteString "<INPUT TYPE=HIDDEN NAME=""method"" VALUE=""" & mapMethod _
          & """>"
    Select Case mapMethod
      Case "single"
        .WriteString "<INPUT TYPE=HIDDEN NAME=""color"" VALUE=""" & mapClr _
          & """>"
        .WriteString "<INPUT TYPE=HIDDEN NAME=""style"" VALUE=""" & mapStyle _
          & """>"
        .WriteString "<H1>Single Value Map</H1>"
      Case "unique"
```

```
    .WriteString "<INPUT TYPE=HIDDEN NAME=""style"" VALUE=""" & mapStyle _
        & """>"
    .WriteString "<INPUT TYPE=HIDDEN NAME=""variable"" VALUE=""" & _
       mapVar  & """>"
    .WriteString "<H1>Unique Value Map of " & mapVar & "</H1>"
   Case "quantiles"
    .WriteString "<INPUT TYPE=HIDDEN NAME=""variable"" VALUE=""" & _
       mapVar & """>"
    .WriteString "<INPUT TYPE=HIDDEN NAME=""classes"" VALUE=""" & _
       mapClasses & """>"
    .WriteString "<INPUT TYPE=HIDDEN NAME=""start"" VALUE=""" & startClr _
        & """>"
    .WriteString "<INPUT TYPE=HIDDEN NAME=""end"" VALUE=""" & endClr _
       & """>"
    .WriteString "<H1>Quantile Map of " & mapVar & "</H1>"
  End Select
  .WriteString "The next mouse click will: "
  .WriteString "<INPUT TYPE=RADIO NAME=""pzi"" VALUE=""identify""> Identify "
  .WriteString "<INPUT SELECTED TYPE=RADIO NAME=""pzi"" VALUE=""zoom"" _
        CHECKED> Zoom And Pan  "
  .WriteString "<SELECT NAME=""level"">"
  For i = 1 To 10
   If i = CInt(mapLevel) Then
    .WriteString "<OPTION SELECTED VALUE=""" & i & """>" & i & "</OPTION>"
   Else
    .WriteString "<OPTION VALUE=""" & i & """>" & i & "</OPTION>"
   End If
  Next
  .WriteString "</SELECT>"
  .WriteString "<BR>"
  'Send the current map extent
  .WriteString "<INPUT TYPE=HIDDEN NAME=""minx"" VALUE=""" & _
        Map1.Extent.Left & """>"
  .WriteString "<INPUT TYPE=HIDDEN NAME=""miny"" VALUE=""" & _
        Map1.Extent.Bottom & """>"
  .WriteString "<INPUT TYPE=HIDDEN NAME=""maxx"" VALUE=""" & _
         Map1.Extent.Right & """>"
  .WriteString "<INPUT TYPE=HIDDEN NAME=""maxy"" VALUE=""" & _
        Map1.Extent.Top & """>"
  If roadsOn Then
   .WriteString "<INPUT TYPE=HIDDEN NAME=""roads"" VALUE=on>"
  End If
  If citiesOn Then
   .WriteString "<INPUT TYPE=HIDDEN NAME=""cities"" VALUE=on>"
  End If
  .WriteString "<TABLE BORDER>"
  .WriteString "<TR><TD>"
  .WriteString "<INPUT TYPE=IMAGE SRC=" & theImage & " > "
  .WriteString "</TD></TR></TABLE>"
  .WriteString "<BR>"
  If mapPzi = "identify" Then
    If gSelection.Count = 0 Then
```

```
      .WriteString "No states were selected"
    Else 'we have at least one selected set.
      Dim tabDesc As New MapObjects2.TableDesc
      Dim curfield As New MapObjects2.Field
      gSelection.MoveFirst
      .WriteString "<TABLE BORDER>"
      Do While Not gSelection.EOF
        Set tabDesc = gSelection.TableDesc
        For i = 0 To tabDesc.FieldCount - 1
          Set curfield = gSelection.Fields(tabDesc.fieldName(i))
          .WriteString "<TR>"
          .WriteString "<TD> " & curfield.Name & "</TD>"
          If curfield.Type = moString Then
            If UCase(curfield.Name) = "STATE_ABBR" Then
              .WriteString "<TD><A HREF=""http://www.state." & _
                curfield.Value & ".us"">" & curfield.Value & "</a>"
            Else
              .WriteString "<TD> " & curfield.Value & "</TD>"
            End If
          Else
            .WriteString "<TD> " & curfield.ValueAsString & "</TD>"
          End If
        Next i 'curfield
        gSelection.MoveNext
      Loop
      .WriteString "</TABLE>"
    End If
  End If
  .WriteString "</FORM></BODY></HTML>"
 End With
End Sub
```

As you can see, this is nearly identical to the version of *MakeWebPage* used in Chapter 14. The first major change is writing out the *ResponseHeader*, which you did in PERL in Chapter 14. The form action is now to call *esrimap.dll*, not the PERL script. The first argument *esrimap.dll* expects is the name of the VB/MO program to which client requests should be passed. Actually, this is a bit misleading. The call is to the name of the service being served by the MOIMS. There can be many different services (VB/MO programs), and they can reside on different machines.

You could have done this in PERL, too. However, you will see in the following that the MOIMS interface makes setting these up fairly painless. The last major change is in the specification of the *jpg* file. In this case, the program uses the file name and path (*theImage*) passed in from *MakeBatchMap*. The other change is quite simple: substituting *.WriteString* for *Print #2*. All that remains is to compile this program and then connect it to the IMS.

Configuring the IMS

To run the MOIMS on your own server, you need either Microsoft or Netscape server software. This may work using other software, but these are the software ESRI recommends. Microsoft's Personal Web Server, which comes with FrontPage, was used for the examples in this chapter. This software sets up a subdirectory for storing web folders. On the author's computer, this subdirectory is named *WebShare*. Under *WebShare*, there is a directory named *scripts*, which houses the DLL *esrimap.dll*.

To start the MOIMS, you need to start three programs: IMS Catalog, IMS Launch, and IMS Administrator. You must start these in this order, and they must be running on the computer that uses the DLL *esrimap.dll*.

NOTE: *A full discussion of these programs, which is beyond the scope of this tutorial, can be found in ESRI's* MapObjects IMS User's Guide.

The following are excerpts of the official descriptions of these programs (and this three-tiered approach) in the *MapObjects IMS User's Guide*: "IMS Catalog serves as a registry for IMS map services that are accessed by Web client applications and IMS Admin. It maintains the registry so Web clients can view services available from the Web site." In short, the catalog stores a list of services (VB/MO programs and ArcExplorer programs).

"IMS Launch, also an in-memory process, prepares a map server to work with MapObjects IMS. IMS Launch runs on each map server computer. IMS Catalog uses IMS Launch to start the requested number of map services on a map server. To IMS-enable a map server computer, all you have to do is install and run IMS Launch." In other words, IMS Launch facilitates communication between the server (where *esrimap.dll* is located) and computers, where the services (the programs and their web addresses in the catalog) are located.

"You use IMS Admin interactively to register and unregister map services with ESRIMap and to maintain the list of current services in the IMS Catalog." The following outlines the processing sequence of ESRI's three-tiered approach.

1. On the *client level*, the client makes a request, perhaps by clicking on a button on a web page. The request goes to the URL of the IMS web site.
2. On the *middleware level*, the request is received at the IMS web site. The server software passes the form inputs to the DLL *ESRIMap.dll*. The DLL "asks" *IMSCatalog* for a list of active services (these may be on different computers). The client's request is then sent to the appropriate service.
3. On the *server level*, the service gets the request, makes the map, and sends messages (the response header and all *WebLink.WriteString* outputs) back to the DLL at the middleware level. On the *middleware level*, *ESRIMap.dll* takes the response from the server and pushes the map (web page) to the client.

If you think about what you covered in Chapter 14, this is behaving much like PERL/VB/HTML actions. However, this should be more efficient because the programs for creating maps are always running.

IMS Admin is where you keep all of the services (VB/MO and ArcExplorer projects) you wish to expose to clients. When IMS Admin starts, it presents an interface for managing services (figure 15-4).

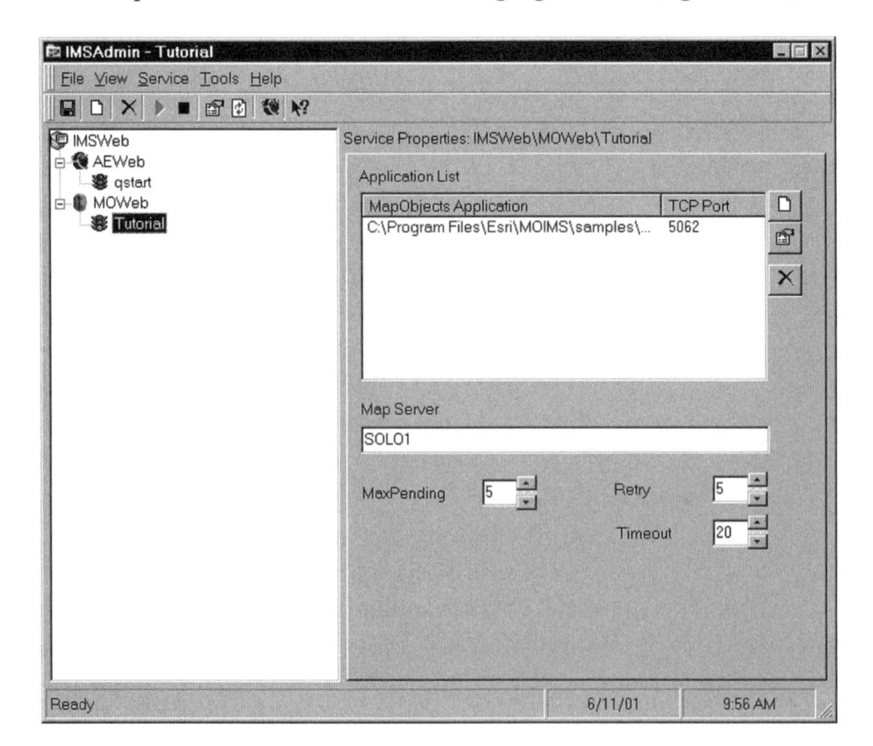

Fig. 15-4. IMS Admin interface.

There are two types of services available: MOWeb (programs written with MapObjects) and AEWeb (ArcExplorer projects). You use the IMS Admin interface to add, delete, start, and stop any of these services. For AEWeb services, you can also specify which files can be downloaded.

Figure 15-5 shows an ArcExplorer project service and the Tutorial service already loaded. In the following, you will load a new service.

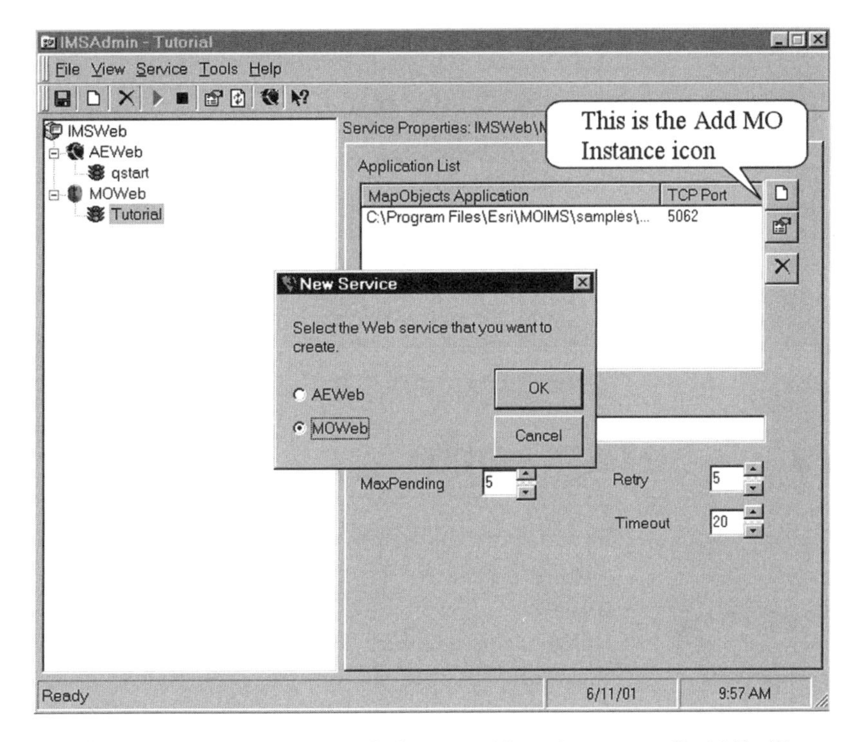

Fig. 15-5. Adding a new service.

1. Click on Service and then on New (or press Ctrl-N). You will be asked to choose the type of service you wish to make (see figure 15-5). Select an MOWeb service.
2. After the service is added, highlight it in the left-hand window (the one with the traffic lights).
3. Click on the AddMO Instance button (the blank sheet at the right) to set the specifics of the new service.

A dialog appears for adding the program, port, and server (see figure 15-6).

4. Set the values as indicated in figure 15-6 and click on OK.

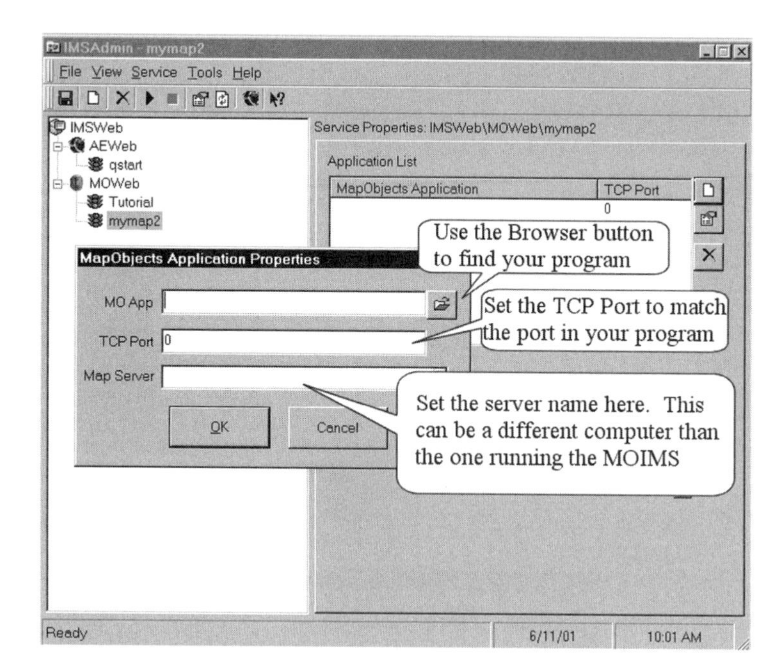

The last step is to start the service.

5 Click on Tools > Start in the IMS Admin main menu, or click on the Start icon (the triangle) on the toolbar.

This will start your program and the traffic light icon will turn from red to green. You are now serving maps with ESRI's MOIMS!

Fig.15-6. Setting the service details.

A Setup Form

In Chapter 14 you used an initialization form (*form1.html*) that allowed users to select a set of map parameters before building the map. There is similar form for the IMS on the companion CD-ROM. It is named *form2.html*. There are only two differences between this and *form1.html*. The first is a cosmetic change. The first line appearing on the web page has been changed from "Serving MapObjects Generated Maps – Method 1" to "Serving MapObjects Generated Maps – Method 2."

The second change deals with calling the proper form action (remember, there are three forms on this page, and each one has a form action). In this page,

```
<form ACTION="http://cowbar2.gg.utk.edu/cgi-bin/method1.pl" METHOD="POST">
```

has been replaced with

```
<form ACTION="http://solo1.gg.utk.edu/scripts/esrimap.dll" METHOD="GET">
 <input type="hidden" name="name" value="Mymap2">
```

These programs were tested on the computer *solo1.gg.utk.edu*. The programs in Chapter 14 were tested on computer *cowbar2.gg.utk.edu*. You will need to change the URL in the form Action line to match your server (the one where the IMS is running).

Server Maintenance

The methods of serving maps presented here and in Chapter 14 both write files to the server. Periodically, you will need to clean out the area at which these files are written. You can either do this by hand (go in and delete them) or create a utility to do this. If you have a systems administrator for your server, he should be able to help with this task.

Comments on the MOIMS

You might wonder if you really need the MOIMS. After all, the PERL script approach in Chapter 14 also worked. However, there are advantages to the approach used in this chapter. First, the VB program is always running. Thus, you do not waste time loading the program into memory and unloading the program after exporting maps. Using VB, you could write a feature similar to the WebLink OCX so that the program would run continuously, but that is beyond the scope of this book.

Another advantage of the MOIMS is that it handles distributed applications quite easily. You could manage multiple sites using PERL. If you are an experienced PERL programmer (or want to become one), you can try this approach.

Perhaps the biggest advantage of the MOIMS is in scalability. The methods outlined in Chapter 14 work for sites that will take a small number of hits (less than one per minute), but for heavy traffic sites, scalability may be a problem. A final advantage of the IMS is that it supports ArcExplorer projects. If that is what you need, the MOIMS is the way to go.

ESRI has recently released ArcIMS. This software comes with an ActiveX control that in many ways is very similar to MO. However, the MO OCX has many more features than the ArcIMS ActiveX control. You can use your MO programs with ArcIMS if you want to take full advantage of MO's capabilities. However, MO does not contain some of the features the ArcIMS ActiveX contains, including translation in ESRI's eXtensible Markup Language, ArcXML. For a more detailed discussion of the similarities and differences between MOIMS and ArcIMS, see the ESRI White Paper "ArcIMS_vs_ MOIMS.pdf."

If you are an accomplished VB/MO programmer, the move to ArcIMS and its ActiveX (via Active Server Pages) should be fairly easy to accommodate. If you want to distribute your existing MO projects over the Web with server-side processing, you can use either MOIMS or ArcIMS. You can use your existing VB/MO applications with ArcIMS. For details on how to serve VB/MO applications with ArcIMS, see the ESRI White Paper "Using MapObjects IMS with ArcIMS."

Summary

In this chapter you used the MOIMS to serve maps over the Internet. This required placing the WebLink control in your program. That control allowed your program to receive requests from the DLL *esrimap.dll* and write responses back to that DLL to be passed on to the client. The process of serving maps was handled by the MOIMS. The program you created for this chapter was nearly identical to that used in Chapter 14. However, using the MOIMS is superior to the PERL script used in Chapter 14. The main advantage of the MOIMS is that your VB/MO program is constantly loaded into memory, as is *esrimap.dll*. This leads to faster response times than the method used in the previous chapter.

GIS software design is a fast-changing field, and web-based GIS is perhaps the fastest changing application in all of GIS. The recent emergence of extensible markup languages (XML) for passing information over the Internet will no doubt cause more change and uncertainty in the GIS community. New XML standards, such as the Open GIS Consortium's Geography Markup Language and ESRI's ArcXML, are evolving rapidly. Nevertheless, for server-side GIS functions based on ActiveX controls, the materials covered in this and the previous chapter will remain relevant. Generating maps for the Web with an ActiveX control, whether it is ESRI's MapObjects or some other vendor's OCX, will require taking inputs from a user, calling the necessary subroutines, and pushing the results back to the client.

Chapter 16

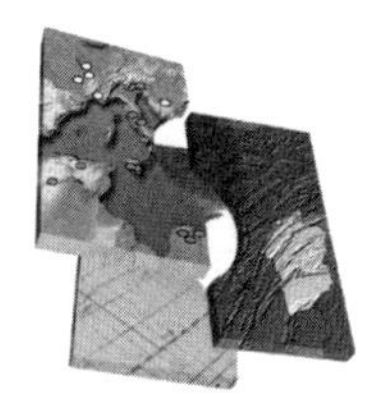

Buffering and Overlay: Part 1

Introduction

The material covered in first fifteen chapters can be implemented with either MapObjects 1.2 or MapObjects 2.0 or later. The following looks at procedures that take advantage of some of the new functionality in MapObjects 2. Starting with MO2, ESRI introduced a buffer method, as well as new methods (such as *intersect, union, xor,* and *difference*) for specifying the relationships among shapes. Another improvement in MO2 was the introduction of coordinate system and projection system objects. These allow for on-the-fly projection and the mixing of map layers that may have different projections or datums. These improvements are discussed in this and the next two chapters.

The project for this chapter introduces a new tool, located on the toolbar, for buffering point layers. Once the layer is buffered, the resultant shapes are intersected with a polygon layer. The goal of this chapter is not only to show how this can be done but to illustrate the dangers of doing geometric operations in non-rectangular coordinate systems. The methods introduced in this chapter are developed further in the following two chapters.

It is important that you use the shapes referenced in this chapter: *knoxtract.shp* and *recycle.shp*. These are located in the *shapes\Knox* directory on the companion CD-ROM. (You could use others, but if they are near the poles or are not in decimal degrees, the methods described in this chapter will not work.) The first shape contains the census tracts taken from the U.S. Bureau of the Census' TIGER 97 files. The second shape, *recycle.shp*, consists of some recycling centers in Knox County, Tennessee. The locations of the centers were geocoded in ArcView 3.

Setting Up the Project

CD-ROM NOTE: *The VB project in the* Chapter16_1 *directory on the companion CD-ROM starts here.*

The first step in building the project is to add a new image to the image list and a new tool to the toolbar on form *Form1*.

1 Add the picture *ellipse.bmp* to the image list.

The image can be found in *C:\Program Files\ESRI\MapObjects2\Samples\VB\MoView2\Bitmaps*. (This assumes that MO2 is installed on the C drive.)

2 Add a new button to the toolbar control. Name the button *PointBuffer*, with a key of *PointBuffer*. The value should be *tbrUnpressed*, and the style *tbrDefault* (see figure 16-1).

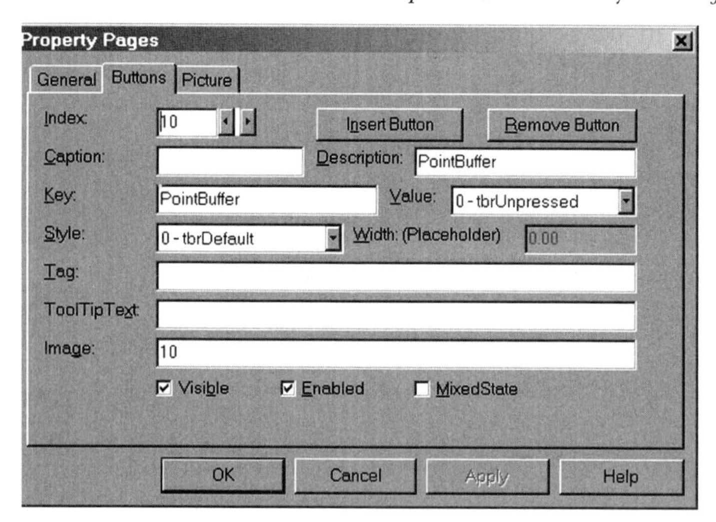

Fig. 16-1. Toolbar properties for the new button.

This new button should work as follows. If there are two or more layers in a map and the active layer is a point layer, the button should be enabled. Otherwise, the button should be disabled. That is, in this chapter you will buffer points only. To control this behavior of the PointBuffer button, you need to add the following lines to the *RefreshCombo1* sub (changes are indicated in bold).

3 Edit the existing *RefreshCombo1* sub to include the lines indicated in bold in the following.

```
Public Sub RefreshCombo1()
 Dim i As Integer
 Dim curselected As String
 Toolbar1.Buttons("PointBuffer").Enabled = False
 If (Combo1.ListIndex >= 0) Then
   curselected = Combo1.List(Combo1.ListIndex)
   Toolbar1.Buttons("Identify").Enabled = True
```

```
   Toolbar1.Buttons("Select").Enabled = True
   cmdSave.Enabled = True
   If (Map1.Layers.Count > 1) Then
   cmdIntersect.Enabled = True
   Else
   cmdIntersect.Enabled = False
   End If
 Else
   curselected = ""
   Toolbar1.Buttons("Identify").Enabled = False
   Toolbar1.Buttons("Select").Enabled = False
   cmdSave.Enabled = False
   cmdIntersect.Enabled = False
 End If
 Combo1.Clear
 If Map1.Layers.Count = 0 Then
   Exit Sub
 End If
 For i = 0 To Map1.Layers.Count - 1
   Combo1.AddItem Map1.Layers(i).Name
   If Combo1.List(i) = curselected Then
     Combo1.ListIndex = i
     Set ActiveLayer = Map1.Layers(i)
     If (ActiveLayer.shapeType = moPoint) And (Map1.Layers.Count >_
        1)  Then
     Toolbar1.Buttons("PointBuffer").Enabled = True
     End If
   End If
 Next i
End Sub
```

The first change, at the beginning of the sub, is for disabling the button. Then, as the content of *Combo1* is rebuilt, the program checks to see if *ActiveLayer* is a point layer and whether there are at least two layers in the map. If so, the button is enabled.

The next task is to figure out what should happen if the user clicks on this button. For now, let's assume that a sub named *BufferPoints* is called. This requires a case in the *Toolbar1_ButtonClick* sub.

4 Edit the *Toolbar1_ButtonClick* sub, making the changes indicated in bold in the following.

```
Private Sub Toolbar1_ButtonClick(ByVal Button As ComctlLib.Button)
 Select Case Button.Key
 Case "LayerControl"
  Form2.Show
```

```
 Case "FullExtent"
  Map1.Extent = Map1.FullExtent
  Map1.Refresh
 Case "PointBuffer"
  Set BufferLayer = ActiveLayer
  BufferPoints
  Map1.Refresh
 End Select
End Sub
```

Here, when the user clicks on the *PointBuffer* button (the one with the ellipse icon), a function named *BufferPoints* will be called. This new function should create the buffers. Once they are created, the map is refreshed in order to draw the buffers. (The *BufferPoints* function is discussed in material to follow.) For now, let's consider the shape of a point buffer.

Point Buffers and Coordinate Systems

Just what should a point buffer be? You may think it should be a circle, but that might not be the case. The shape of a buffer is dependent on the map coordinate system. If the coordinate system is rectangular, circular-shaped point buffers should result. However, if the map units are decimal degrees, circular-shaped buffers are not correct.

The following example illustrates this difference. Using ArcView 3.2, two views of the same area (Knox County, Tennessee) were constructed. The first view was in decimal degrees. The second was in state plane coordinates (feet). (The Tennessee 1983 state plane projection is a Lambert Projection, discussed further in Chapter 17.) Equal-size buffers were then constructed in both views, using ArcView 3.2's buffering function. Figure 16-2 shows the buffers in decimal degrees, whereas figure 16-3 shows the same map but in state plane coordinates.

The resulting buffer shape from the rectangular coordinate system was then pasted into the decimal degree view. ArcView, knowing the incoming shape had its own projection system, projected it on the fly to accurately depict it in decimal degrees. The result is shown in figure 16-4.

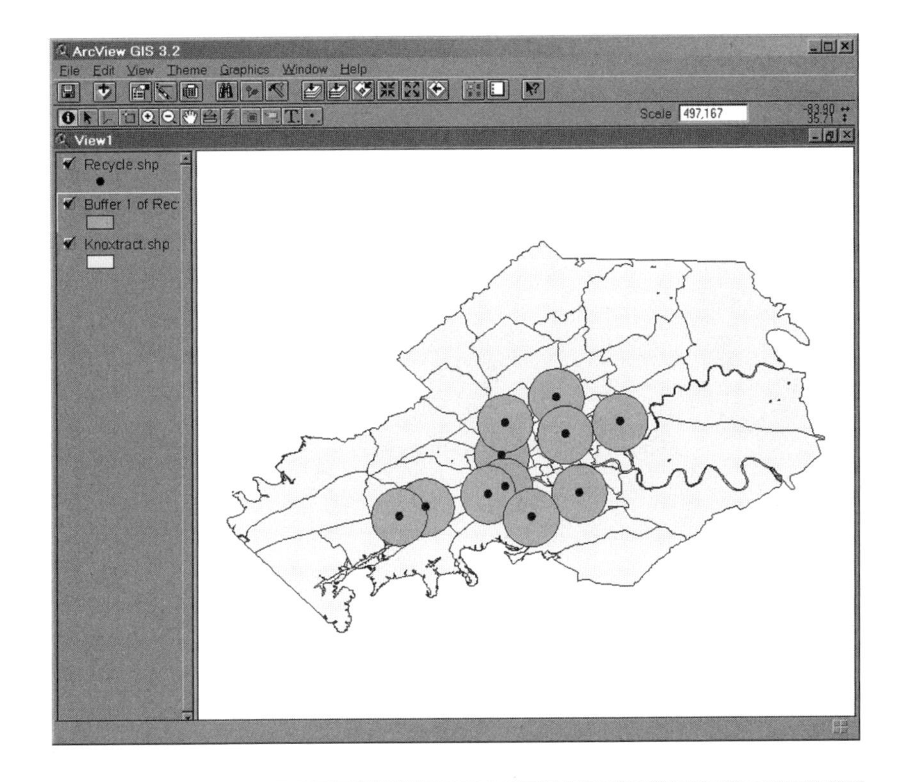

Fig. 16-2. Decimal degree view.

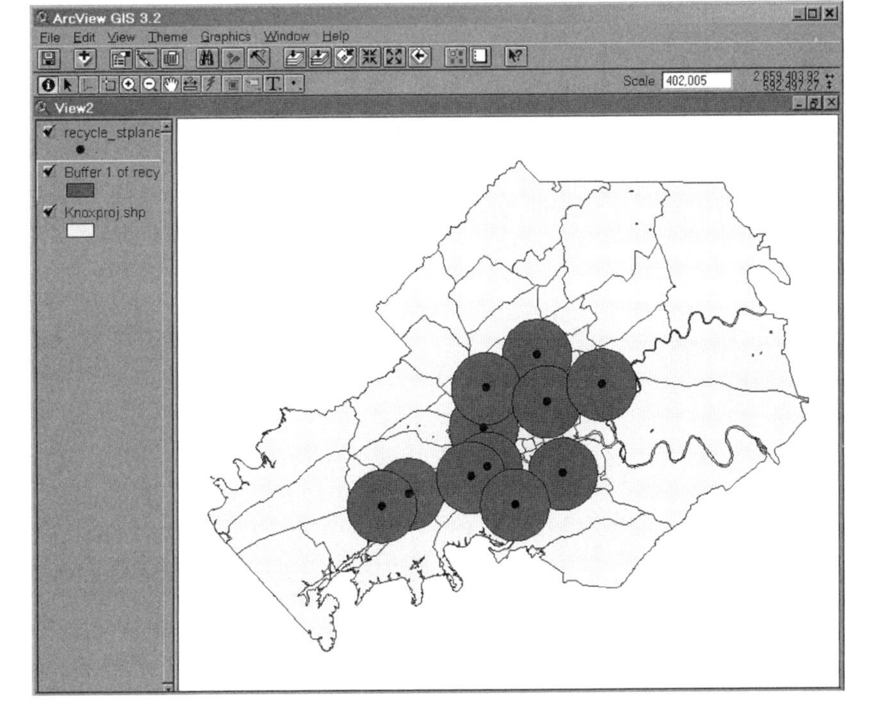

Fig. 16-3. State plane coordinate view.

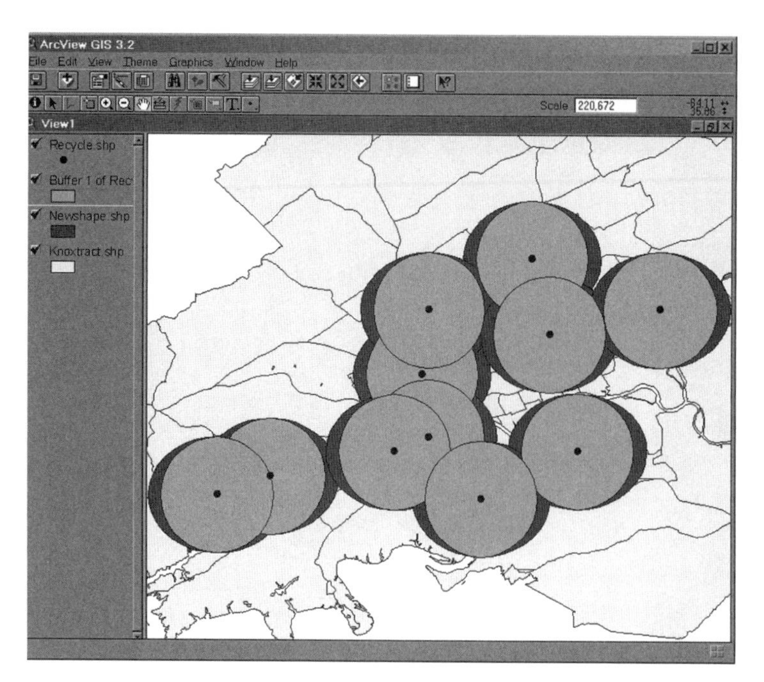

Fig. 16-4. Comparison of the two sets of buffers.

Look at what happened. The buffers generated in the state plane system appear as ellipses when projected into the decimal degree view. What ArcView has done in the decimal degree view is to construct circles when it should have constructed ellipses. That is, the buffers in the decimal degree view are approximate. If the buffer sizes are small or if you are near the equator, this is not a bad approximation. However, if the buffers are large or you are near the poles, this can be very inaccurate.

The reason for the changing shape of a point buffer when displayed in decimal degrees is that the distance covered by one degree of latitude does not equal the distance covered by one degree of longitude, except at the equator. One degree of latitude equals approximately 69.17 miles, depending on the datum used (discussed further in Chapter 17). At the equator, 1 degree of longitude is also 69.17 miles. At the poles, one degree of longitude covers 0 miles. At 60 degrees, one degree of longitude is approximately 34.59 miles in length. In general, at any latitude, the distance represented by one degree of longitude is:

Distance of 1 degree longitude = 69.17 * cos(latitude)

Knox County is at approximately 36 degrees north latitude. The cosine of 36 degrees is 0.809, meaning that it takes 1.235 degrees of longitude to equal the same distance as 1 degree of latitude. Thus, a two-mile buffer will have a north-south diameter of 0.029 degrees of latitude, but an east-west diameter of approximately 0.035 degrees of longitude. This is why the buffers in figure 16-4 are elliptical.

Let's try a different approach from circles. In the following you will construct elliptical buffers in a decimal degree map. For mid latitude ranges, ellipses are good approximations of point buffers. Near the poles they are not.

NOTE: *The method used in this chapter assumes the incoming map layers are in decimal degrees and that you are not too close to the poles. This issue is addressed more thoroughly in the next two chapters.*

■■ The BufferPoints Sub

You have seen that for mid to low latitude areas and small buffer distances, an ellipse is the preferred point buffer shape when working in decimal degrees. You will develop a function that will generate elliptical buffers. However, as you build these new shapes, you will need to keep them in a collection. Therefore, you need to add the following declarations to the top of form *Form1*'s code page.

1 Enter the following.

```
Dim colEllipseRect As New Collection
Public bolBufferOverlay As Boolean
Public BufferLayer as MapObjects2.MapLayer
```

The collection *colEllipseRect* will store all buffers you create. *BufferLayer* is the point layer that is buffered. The Boolean will keep track of whether or not you are buffering.

Now you are ready to write the *BufferPoints* sub. For now, assume that you are building a two-mile buffer. (This assumption will be relaxed later.)

2 Enter the following in the form *Form1*'s code page.

```
Public Sub BufferPoints()
 Dim buffsize As Double
 Dim recs As New MapObjects2.Recordset
 Dim pt As New MapObjects2.Point
 Dim lyr As MapObjects2.MapLayer
 Dim xoffset, yoffset As Double
 Dim radians, pi As Double
 Dim degpermile As Double
 Dim i As Integer
 pi = 3.14159265358979
```

```
 degpermile = 0.0144568
 buffsize = 2
 'Clear out the collection if we are re-buffering
 If colEllipseRect.Count > 0 Then
  For i = 1 To colEllipseRect.Count
  colEllipseRect.Remove (1)
  Next
 End If
 Set recs = BufferLayer.Records
 recs.MoveFirst
 Do While Not recs.EOF
  Dim EllipseRect As New MapObjects2.Rectangle
  Set EllipseRect = Nothing
  Set pt = recs.Fields("Shape").Value
  yoffset = degpermile * buffsize
  radians = Abs(pt.Y * pi / 180#)
  xoffset = degpermile * buffsize / Cos(radians)
  EllipseRect.Bottom = pt.Y - yoffset
  EllipseRect.Top = pt.Y + yoffset
  EllipseRect.Left = pt.X - xoffset
  EllipseRect.Right = pt.X + xoffset
  colEllipseRect.Add EllipseRect
  recs.MoveNext
 Loop
 bolBufferOverlay = True
End Sub
```

This sub begins by declaring variables you will need. It then defines some values. The meaning of *Pi* should be obvious. *degpermile* is the number of degrees per mile of latitude. Put another way, one degree of latitude equals approximately 69.17 miles. The buffer distance is 2 miles.

The program then checks to see if it should be re-buffering. If so, it removes all content from the *colEllipseRect* collection. It then gets the *BufferLayer*'s records, moves to the first record, and then loops through all records. Let's look at the heart of the loop.

```
Dim EllipseRect As New MapObjects2.Rectangle
Set EllipseRect = Nothing
Set pt = recs.Fields("Shape").Value
yoffset = degpermile * buffsize
radians = Abs(pt.Y * pi / 180#)
xoffset = degpermile * buffsize / Cos(radians)
EllipseRect.Bottom = pt.Y - yoffset
EllipseRect.Top = pt.Y + yoffset
EllipseRect.Left = pt.X - xoffset
```

```
 EllipseRect.Right = pt.X + xoffset
 colEllipseRect.Add EllipseRect
```

This loop begins by creating a new rectangle to hold the bounding box of the ellipse that corresponds to the current element in the buffer layer. That is set to *Nothing*. Next, the program gets the point corresponding to the current record. The program determines a *yoffset* (the change in latitude) by multiplying the buffer size, in miles, by the degrees per mile. Before determining the *xoffset* (the change in longitude), the program has to convert the point's latitude value to radians (the cosine function requires that angles be expressed in radians). Once the offset values are obtained, it is a simple matter to set the bounding box of the ellipse. The resulting rectangle is then added to the collection *colEllipseRect*.

Drawing the Buffers

The buffers will be drawn after all layers are drawn. Thus, the following is added to the end of the *AfterLayerDraw* sub.

1 Enter the following.

```
'This goes at the end of the AfterLayerDraw sub
 If index = 0 And bolBufferOverlay Then
  Dim poly1, poly2 As New MapObjects2.Polygon
  Dim el As New MapObjects2.Ellipse
  Dim buffersym As New MapObjects2.symbol
  buffersym.Color = moRed
  buffersym.Style = moLightGrayFill
  For i = 1 To colEllipseRect.Count
  With colEllipseRect.Item(i)
   el.Bottom = .Bottom
   el.Top = .Top
   el.Left = .Left
   el.Right = .Right
  End With
  If i = 1 Then
   Set poly1 = el
  End If
  Set poly2 = poly1.Union(el)
  Set poly1 = poly2
 Next
 Map1.DrawShape poly2, buffersym
End If
```

This section of the sub begins by checking if the top layer has been drawn (*index* = *0*) and if buffering is being performed. If so, the program declares several objects that will be needed. The draw symbol is set to red, with a light gray fill. This will allow lower layers to show through the buffers.

A new ellipse object, *el*, is created for every element of the collection of bounding rectangles. The program then finds the union of all ellipses. When all elements have been processed, the program draws the new shape (*poly2*) with the proper symbol. The result should look like that shown in figure 16-5.

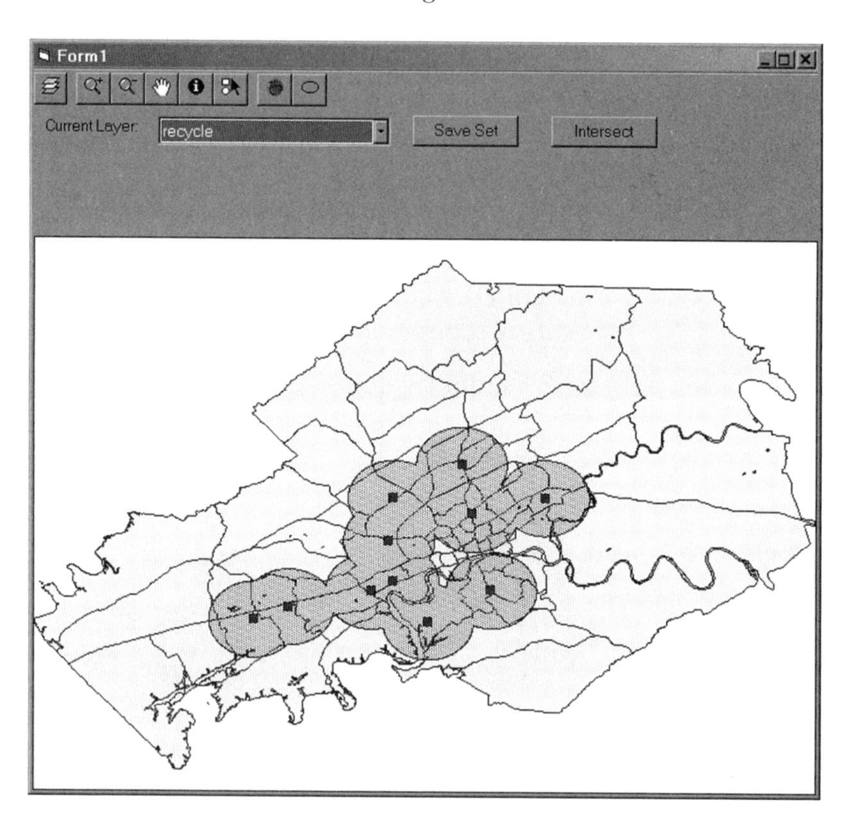

Fig. 16-5. Buffered recycling centers.

Figure 16-5 was made possible by a new method in MO2, *Union*. *Union* is one of several new methods that can be applied to points, lines, and polygons. Other methods include *Intersect*, *Difference*, and *Xor*. You will use the *Intersect* method in the next section.

▪▪ Specifying Overlay Parameters

CD-ROM NOTE: *The VB project in the* Chapter16_2 *directory on the companion CD-ROM starts here.*

Seeing buffers is nice, but you typically want to use them to select features or to perform some other overlay function. To make your program more robust, you will allow the user to select the overlay layer (which for now you can assume is a polygon layer) and a summation variable. You will estimate the value of the summation variable in a buffer by area weighting.

Suppose, for example, that the summation variable is total population. If 40% of a polygon's area lies within a buffer, you would estimate the value of the summation variable for the intersection of the buffer and the polygon in question to be 40% of the population in that polygon. This is based on the assumption of even distribution of population, an assumption that may not be correct. Nevertheless, this approach—area weighting—is very common in GIS studies. In order to implement this approach, you need to add the following declaration to the top of form *Form1*'s code page (the new lines are in bold).

1 Enter the following indicated in bold.

```
Public colRecSetClass As New Collection
Public bolBufferOverlay As Boolean
Dim colEllipseRect As New Collection
Public BufferLayer As MapObjects2.MapLayer
Public OverlayLayer As MapObjects2.MapLayer
Public SumVariable As MapObjects2.Field
Public BufferDistance As String
```

When the user clicks on the Buffer icon (the ellipse in the toolbar), a form will appear that will allow the user to set the overlay layer, the summation variable, and the buffer distance. To incorporate this form, you need to add the following lines to the *Toolbar1_ButtonClick* event. (The changes are in bold.)

2 Enter the following indicated in bold.

```
Case "PointBuffer"
 frmBuffer.Show vbModal
 If bolBufferOverlay Then
 Set BufferLayer = ActiveLayer
 BufferPoints
 Map1.Refresh
```

```
            frmReport.Show
        End If
```

A click on the Point Buffer tool will display a new form, *frmBuffer*. This form should be modal. That is, the user must complete the form or dismiss it before any other actions can take place. If the forms for specifying the overlay layer, summation distance, and buffer distance are filled out successfully, *bolBufferOverlay* will be set to True. After the points are buffered, the program will present a report form of the overlay analysis.

The next task is to create the necessary forms. Let's begin with the *frmBuffer* form.

3 Create a form that looks like that shown in figure 16-6. Incorporate a label for storing the name of the layer to be buffered (*Label2* in figure 16-6), two combo boxes (one for choosing an overlay layer, the other for setting the buffer distance), and two command buttons (*cmdCancel* and *cmdNext*).

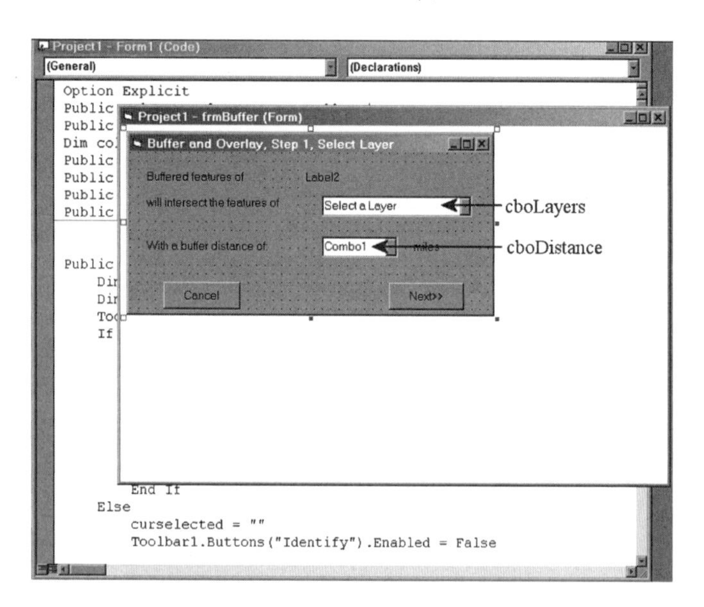

Fig. 16-6. The frmBuffer *form.*

Label2 should be set to the name of the active layer, the *cboLayers* com-bo box should be seeded with a list of available layers, and *cboDistance* should contain a list of reasonable buf-fer distances. (You might want to try replacing this with a slider control.)

You initialize the form by setting the value of *Label2* and populating the combo boxes.

4 Enter the following in the form *frmBuffer* code page.

```
Private Sub Form_Load()
'This is frmBuffer's load function
 Dim i As Integer
 For i = 0 To Form1.Map1.Layers.Count - 1
  If ActiveLayer.Name <> Form1.Map1.Layers(i).Name Then
```

```
  cboLayers.AddItem Form1.Map1.Layers(i).Name
  End If
 Next
 cboLayers.ListIndex = 0
 For i = 1 To 10
  cboDistance.AddItem i * 0.5
 Next
  cboDistance.ListIndex = 4
 Label2.Caption = ActiveLayer.Name
End Sub
```

If the user clicks on Cancel, the program needs to unload the form and set the Boolean for buffering to False.

5 Enter the following.

```
Private Sub cmdCancel_Click()
 Unload Me
 Form1.bolBufferOverlay = False
End Sub
```

If the user clicks on Next, the program needs to store the values in the form to the appropriate variables (that belong to form *Form1*), and display another form for determining the summation variable.

6 Enter the following.

```
Private Sub cmdNext_Click()
 Dim i As Integer
 For i = 0 To Form1.Map1.Layers.Count - 1
 'We must loop through all cases because the entries in cboLayers do NOT
 'match the map layers collection
   If Form1.Map1.Layers(i).Name = cboLayers.List(cboLayers.ListIndex) Then
   Set Form1.OverlayLayer = Form1.Map1.Layers(i)
   End If
 Next
 Form1.BufferDistance = cboDistance.List(cboDistance.ListIndex)
 frmSumVariable.Show vbModal
 Unload Me
End Sub
```

The next task is to create the form for specifying the summation variable. This form will contain a combo box that lists all numeric variables in the overlay layer's attribute file. It will also contain Cancel and Finish buttons. Figure 16-7 shows such a form.

7 Create a form that looks like that shown in figure 16-7. Name the form *frmSumVaraible*. Name the combo box *cboVariables*.

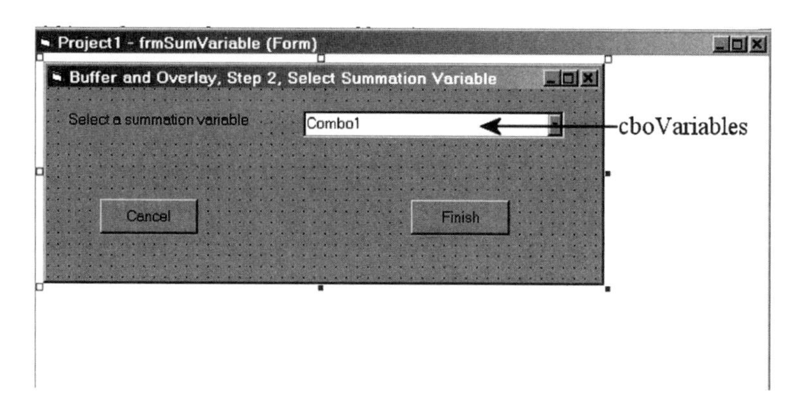

Fig. 16-7. The frmSumVariable *form.*

You initialize the form by seeding the combo box *cboVariables* with the numeric variables, except for the feature ID.

8 Enter the following in the form *frmSumVariable* code page.

```
Private Sub Form_Load()
 Dim recs As New MapObjects2.Recordset
 Dim i As Integer
 Dim fld As MapObjects2.Field
 Set recs = Form1.OverlayLayer.Records
 For Each fld In recs.Fields
  If fld.Type = moLong Or fld.Type = moDouble Then
  If fld.Name <> "FeatureId" Then
    cboVariables.AddItem fld.Name
    End If
  End If
 Next
 cboVariables.ListIndex = 0
End Sub
```

If the user clicks on Finish, the choice is saved. When the form is unloaded, control passes back to the *cmdNext_Click* sub in *frmBuffer.*

9 Enter the following for the *cmdFinish* button's click event on form *frmSumVariable.*

```
Private Sub cmdFinish_Click()
 Dim recs As New MapObjects2.Recordset
 Dim i As Integer
 Dim fld As MapObjects2.Field
 Set recs = Form1.OverlayLayer.Records
 For Each fld In recs.Fields
  If fld.Name = cboVariables.List(cboVariables.ListIndex) Then
  Set Form1.SumVariable = fld
  End If
```

```
 Next
 Form1.bolBufferOverlay = True
 Unload Me
End Sub
```

Finally, you need to handle the case where the user clicks on Cancel.

```
Private Sub cmdCancel_Click()
 Form1.bolBufferOverlay = False
 Unload Me
End Sub
```

There is one more form you need to create—a form to report the results of the buffer and overlay analysis. Create a form named *frmReport.* This is a very simple form, containing only a large text box. Be sure the form's properties are set so that scroll bars are enabled and that *multiline* is set to True (figure 16-8).

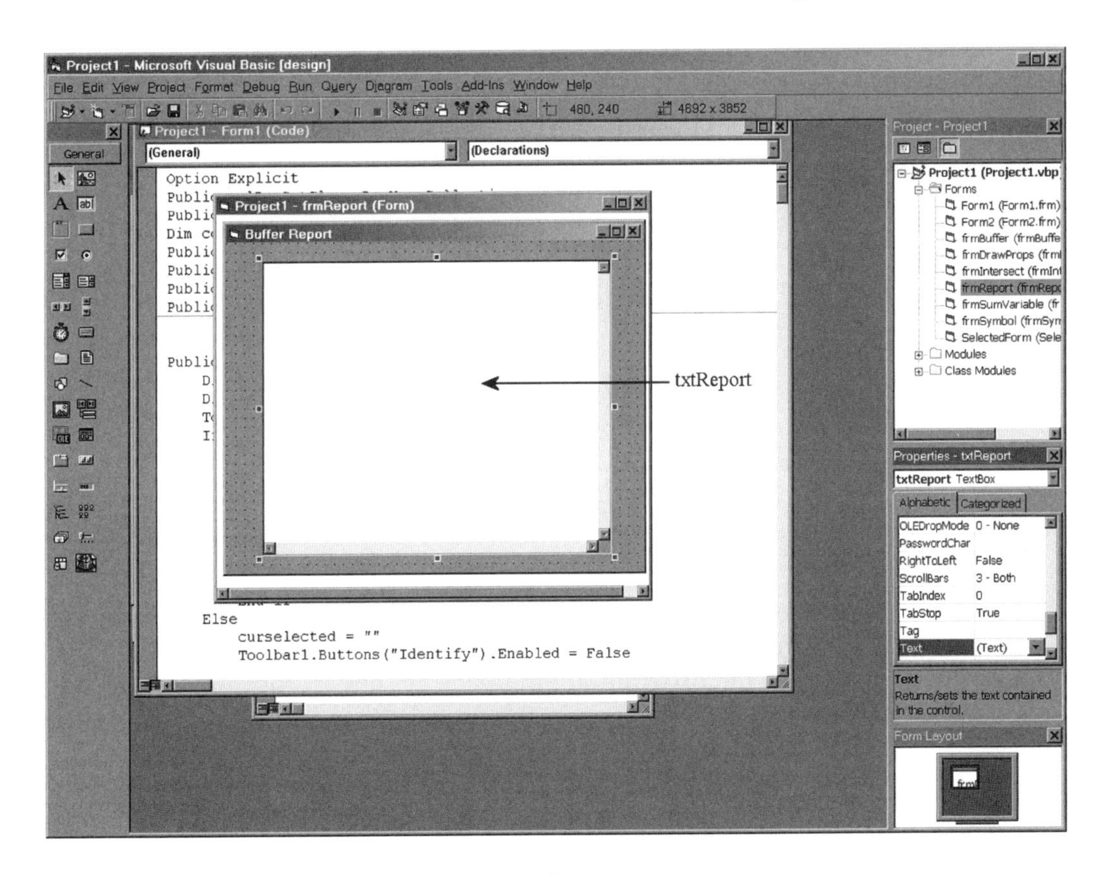

Fig. 16-8. The frmReport *form.*

Now that the buffer specification and report forms are written, you can make the necessary changes in form *Form1*'s *BufferPoints* and *AfterLayerDraw* subs.

The Intersect Function

Now that the buffer and overlay parameters are known, you can implement the *Intersect* function and estimate the value of the summation variable inside the buffers. Make the following changes to the *BufferPoints* sub in form *Form1*'s code page.

1. Remove the following line from the end of the sub.

   ```
   bolBufferOverlay = True
   ```

2. Replace

   ```
   buffsize = 2
   ```

   with

   ```
   buffsize = Cdbl(BufferDistance)
   ```

The major changes for implementing the *Intersect* function lie in intersecting each buffer with the overlay layer and estimating the new value of the summation variable. These steps are done in the *AfterLayerDraw* sub. You begin by declaring a new string variable, *reptext.* This is used to store the text that will appear in the report form's text box.

3. In *AfterLayerDraw,* find the line *buffersym.Style = moLightGrayFill.* Enter the following lines.

```
Dim reptext As String
reptext = "The summation variable is " & SumVariable.Name & vbCrLf
reptext = reptext + "The buffer distance is " & BufferDistance & _
   " miles" & vbCrLf & vbCrLf
```

The next section of the sub (defining each ellipse's bounding rectangle) remains unchanged until after you set *poly1* equal to *poly2.* The program then needs to check for the intersection between each polygon in the overlay layer and the current buffer ellipse. If they do intersect, the program needs to weight the polygon's value on the summation variable by the proportion of the polygon that lies in the ellipse buffer.

4. In *AfterLayerDraw,* find the line *Set poly1 = poly2.* Enter the following lines before the *Next* line.

```
Dim curpoly As New MapObjects2.Polygon
Dim selpoly As Object
Dim allrecs As MapObjects2.Recordset
Set allrecs = OverlayLayer.Records
allrecs.MoveFirst
Dim popest As Double
popest = 0
Do While Not allrecs.EOF
  Set curpoly = allrecs("Shape").Value
  'Note the use of MO2's Intersect method
  Set selpoly = curpoly.Intersect(el)
  If Not selpoly Is Nothing Then
    popest = popest + allrecs(SumVariable.Name) * selpoly.Area / _
      curpoly.Area
    End If
    allrecs.MoveNext
  Loop
  reptext = reptext + "The estimated value within the buffer of " & _
      buffrecs.Fields("Storename").Value & " is " & Format(popest, _
      "######0") & vbCrLf
  buffrecs.MoveNext
```

Finally, the program needs to estimate the value of the population inside the union of all buffers. Note that you cannot simply add up the population estimates made previously because of possible overlaps in the buffers. This is where MO2's *Union* method is so helpful.

5. Edit the remainder of the *AfterLayerDraw* sub, making the changes indicated in bold in the following.

```
Map1.DrawShape poly2, buffersym
popest = 0
allrecs.MoveFirst
Do While Not allrecs.EOF
  Set curpoly = allrecs("Shape").Value
  Set selpoly = curpoly.Intersect(poly2)
  If Not selpoly Is Nothing Then
    popest = popest + allrecs(SumVariable.Name) * selpoly.Area / _
        curpoly.Area
  End If
  allrecs.MoveNext
Loop
reptext = reptext + vbCrLf + "The estimated value within all _
       buffers is " &  Format(popest, "######0")
frmReport.txtReport = reptext
End If
```

When executed, you should get a result like that shown in figure 16-9.

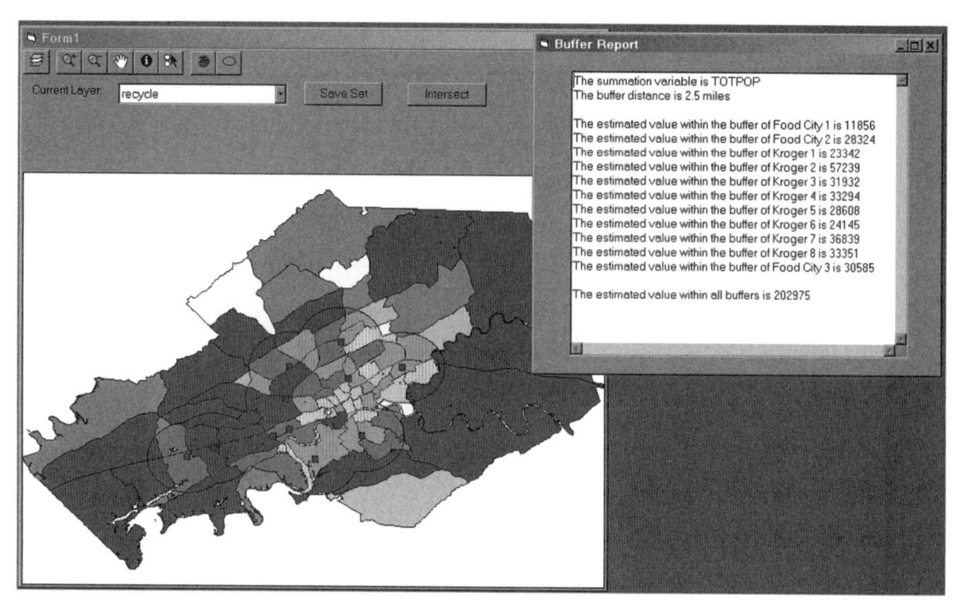

Fig. 16-9. Buffer and Intersect report.

Summary

In this chapter you developed methods for buffering points and intersecting the resulting buffers with polygons. Using area weighting you were able to estimate the value of a summation variable within the buffers. MO's *Intersect* method allowed you to calculate the percentage of a polygon's area that lies within another polygon's area.

The *Union* method allowed you to create a shape that is a union of shapes. The resultant shape, when intersected with a polygon layer, allowed you to calculate the percentage of each polygon in that layer within the union shape. This eliminated the possibility of double counting areas where the buffer polygons overlapped. You also saw how the shape of a point buffer is dependent on the coordinate system of the shapes you are using. To compensate for decimal degrees being spherical (not rectangular) coordinates, you created elliptical point buffers.

The current project works, but it is very limited. For example, you are limited to buffering points and overlaying polygons. These limitations are addressed in Chapter 18. However, a more serious limitation may well be the use of decimal degrees in the computation of areas. You saw how the shape of a buffer changes when projected into decimal degrees, and you compensated for this by using ellipses for the buffer areas.

However, even if buffer shapes are more accurate, it does not follow that area calculations will be accurate. This is particularly true near the poles. If your incoming data were all in projected units, and each layer used the same projection, the area calculations would be acceptable. However, you may often work with data that uses different projections or different datums.

To overcome these problems, you need to consider map projections. If you can work in projected space, the shape of buffers and area calculations will be more accurate. Furthermore, you can take advantage of MapObjects 2's new *Buffer* method. (You can use this on latitude/longitude, but MO assumes you are giving it rectangular coordinates, just as ArcView did in figure 16-2.) In the next chapter you will work with MO2's new projection-on-the-fly capabilities. Once you have those in your project, you can return to buffering and overcome many of the shortcomings of the current approach.

Chapter 17

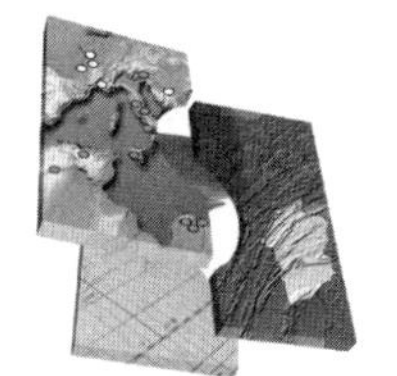

On-the-Fly Projections

Introduction

We all know that the earth is not flat. However, when we draw maps we try to make it appear so. This can cause problems, as you saw in the previous chapter. Latitude and longitude are not rectangular coordinates, and when we treat them as such we make mistakes in our area and length calculations (see figure 16-4, for example).

Fortunately, one of the new features in MO2 helps you deal with this problem. You can now mix data sources that have different datums and projections (defined in material to follow), make sure they register, and present them in a single, consistent coordinate system. Before delving into the details of how MO deals with such issues, it is useful to review some aspects of how location is measured on the earth.

Datums and Geographic Coordinates

You may think that the earth is a sphere or a ball, but this is not true. The earth is not symmetric about its center. That is, there is no single radius that defines the shape of the earth. A closer approximation—but still an approximation—is to say the earth is a spheroid. The difference between a sphere and a spheroid is that a sphere is generated by rotating a circle about an axis, whereas a spheroid is generated by rotating an ellipse about an axis.

An ellipse has major and minor axes, and spheroids are said to have semi-major and semi-minor axes. If you go the MO2 Help file and look up "Spheroid Object" and then click on the word *Spheroid*, you can find a diagram of these axes. The polar axis is the semi-minor axis, whereas the equatorial axis is the semi-major axis. That is, the equatorial axis, the distance from the center of the earth to the equator, is greater (longer) than the polar axis, the distance from the center of the earth to the equator. Put another way, you can think of the earth as a sphere with middle-aged spread!

Throughout cartographic history, there have been attempts to measure the axes of the spheroid. (For a nice discussion, see *Geodesy for the Layman*, available at the National Imagery and Mapping Agency's web site, *http://164.214.2.59/publications/pub.html*). These different measurements give rise to different datums. As technology has improved, our ability to measure the semi-major and semi-minor axes of the earth has improved. However, changes in technology are not the only reason we have many different datums. As mentioned previously, a spheroid is an approximation of the earth's shape.

Not surprisingly, the approximation used depends on the location for which we are trying to measure latitude and longitude. If you think about it (and if you study pictures of the earth's shape), the estimation of the best spheroid for measuring latitude and longitude for North America might be different than that, say, for South America. A datum for the entire world will necessarily have some compromise. Put another way, any world datum will be more accurate in some places than in others.

The result is that there are many different datums you can use. Thus, even if you are given two shape files of the same area in the same units (latitude and longitude) and at the same scale, they will not match if they are based on different datums. In the previous chapter, you used shapes of Knox County, Tennessee, that were extracted from the U.S. Census Bureau's TIGER 97 files.

Since TIGER 94, the Census Bureau has used North American Datum 1983 (often called NAD83 for short) for estimating latitude and longitude. Prior to TIGER 94, latitude and longitude in TIGER files were based on North American Datum 1927. Thus, a TIGER 94 and TIGER 92 map of the same place may not coincide. Table 17-1 provides the differences between the axes for the two datums (to the nearest meter).

Table 17-2: Axis Differences Between NAD27 and NAD83

Axis	NAD27 (m)	NAD83 (m)
Semi-major axis	6378206	6378137
Semi-minor axis	6356584	6356752

There does not appear to be much difference between the NAD27 and NAD83 measures, but when working on a local scale, these differences can be quite apparent. MapObjects 2 uses *GeographicCoordinateSystem* (GCS) objects to keep track of the datums that define latitude and longitude. MO2 comes with 246 predefined Geographic Coordinate System constants (values the *GeographicCoordinateSystem* object can take). If you look in the Help file, you can get a list of these (see figure 17-1).

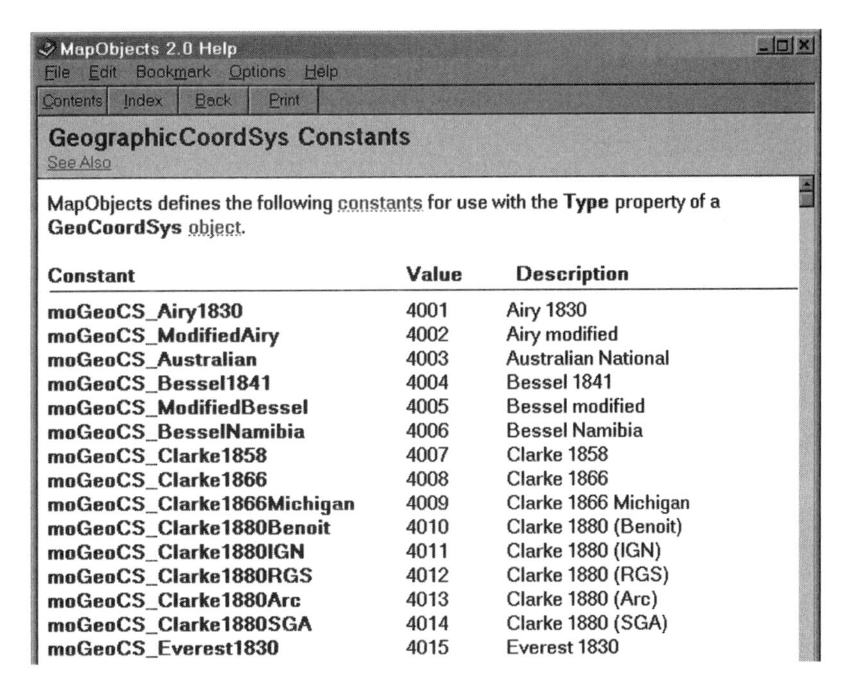

Constant	Value	Description
moGeoCS_Airy1830	4001	Airy 1830
moGeoCS_ModifiedAiry	4002	Airy modified
moGeoCS_Australian	4003	Australian National
moGeoCS_Bessel1841	4004	Bessel 1841
moGeoCS_ModifiedBessel	4005	Bessel modified
moGeoCS_BesselNamibia	4006	Bessel Namibia
moGeoCS_Clarke1858	4007	Clarke 1858
moGeoCS_Clarke1866	4008	Clarke 1866
moGeoCS_Clarke1866Michigan	4009	Clarke 1866 Michigan
moGeoCS_Clarke1880Benoit	4010	Clarke 1880 (Benoit)
moGeoCS_Clarke1880IGN	4011	Clarke 1880 (IGN)
moGeoCS_Clarke1880RGS	4012	Clarke 1880 (RGS)
moGeoCS_Clarke1880Arc	4013	Clarke 1880 (Arc)
moGeoCS_Clarke1880SGA	4014	Clarke 1880 (SGA)
moGeoCS_Everest1830	4015	Everest 1830

Fig. 17-1. Part of the GCS enumeration.

Each datum corresponds to a *datum* object. A *datum* object contains a *spheroid* object, which has an axis property and a flattening property, among others. These are depicted in figure 17-2, which shows a section of the MO object diagram.

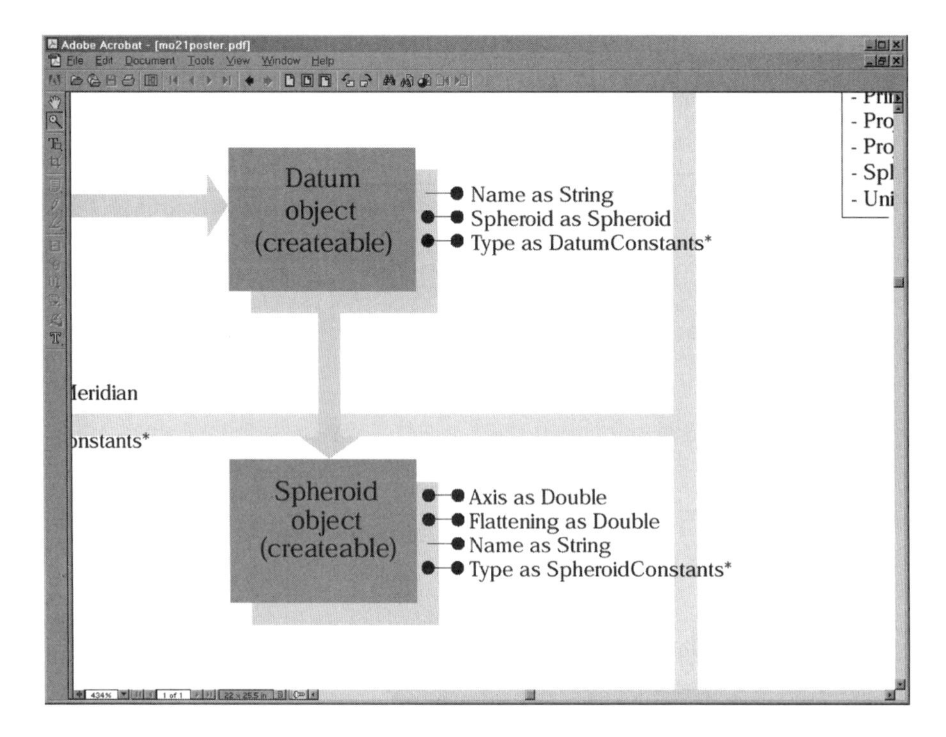

Fig. 17-2. Datum *and* spheroid *objects.*

The axis of a spheroid returns the semi-major axis. The flattening coefficient is equal to $(a - b)/a$, where a is the semi-major axis and b is the semi-minor axis. Let's write some code for returning the axes for a particular geographic coordinate system.

CD-ROM NOTE: *The VB project in the* Chapter17_1 *directory on the companion CD-ROM starts here.*

1. Start a new project and add the MO2 component. On the main form, create a map control, two label boxes, and a combo box. Name the combo box *cboGCS*. Figure 17-3 shows the completed form.

You will not need a large map control, because you will not be drawing any maps in this project. However, you do need a control on the screen, because you will want to access some MO2 objects.

In this project you will present the user with a list of available *GeoCoordSys* objects (geographic coordinate system, or latitude and longitude). The user will choose one, and your program will use the chosen GCS to calculate the semi-minor axis. It will then report the semi-major and semi-minor axes.

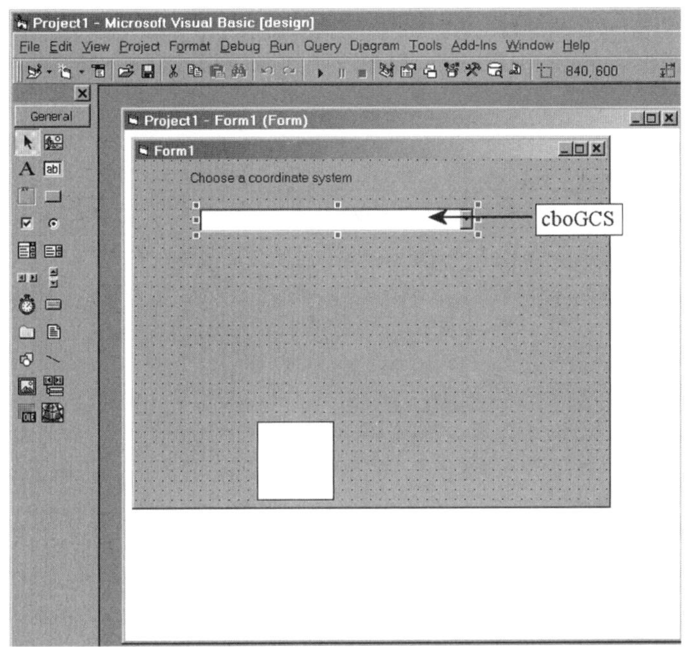

Fig. 17-3. Form layout.

The list of GCSs available will be read in using an MO function called *PopulateWithGeographicCoordSys*. This is a function that loads all available GCS enumeration values (the items in figure 17-1) into an MO2 *Strings* object. Before proceeding to the code, it is worth checking to see if you have MO2 or MO2a (or later). The later upgrade makes loading the *GeographicCoordSys* method much faster. In the first release of MO2, this method (and its sister method, *PopulateWithProjections*) runs much slower than in the later release. The MO2 upgrade notes state:

> Projection performance improvements: Poor performance could be experienced when reading *.prj* files, or using the *PopulateWith . . .* methods on the Strings Collection Object. This was due to additional checks being made for the NADCON grid files (located in the *PE_Grids* folder in *C:\Program Files\Common Files\ESRI*). Now, at 2.0a, if this folder is not present then these checks are not made and performance may be greatly improved. Therefore, if NADCON support is not required for projections then the *PE_Grids* folder should be renamed or removed.

If you do not need NADCON grids, rename that folder. There will be a noticeable improvement in execution speed. Continue with the following, which creates a *Strings* object to hold the elements of the GCS enumeration.

2 Enter the following.

```
Option Explicit
Public stringsGCS As New MapObjects2.Strings
```

The *Form_Load* function then needs to populate the combo box with all elements in the *Strings* object.

3 Enter the following.

```
Private Sub Form_Load()
  Dim i As Integer
  stringsGCS.PopulateWithGeographicCoordSys
  For i = 0 To stringsGCS.Count - 1
    cboGCS.AddItem stringsGCS.Item(i)
  Next
End Sub
```

The *cboGCS_Click* sub is where the user's choice of GCS is processed. The following defines some necessary variables.

4 Enter the following.

```
Private Sub cboGCS_Click()
  Dim geoCoordSys As New MapObjects2.geoCoord-
Sys
  Dim fullstring As String
  Dim strvalue As String
  Dim token As Integer
  Dim value As Long
  Dim flattening As Double
  Dim minoraxis As Double
  Dim majoraxis As Double
```

The program then needs to get the user's choice and extract the numeric part, converting it to a long integer.

5 Enter the following.

```
'Get the user choice and parse it
fullstring = cboGCS.List(cboGCS.ListIndex)
token = InStr(fullstring, "[")
strvalue = Right(fullstring, Len(fullstring) - token)
value = CLng(Left(strvalue, Len(strvalue) - 1))
```

Thus, if the user chose *GCS_North_American_1983 [4269]*, the value would be set to 4269. The following initializes the *geoCoordSys* variable to *Nothing* and then sets its type to *value.* The program then gets the datum for that type of coordinate system, and then the sphere for that datum. Finally, it gets the flattening and axis values, from which it calculates the minor axis.

6 Enter the following.

```
Set geoCoordSys = Nothing
geoCoordSys.Type = value
Dim sphr As New MapObjects2.Spheroid
Set sphr = geoCoordSys.Datum.Spheroid
```

```
  Label2.Caption = "Major Axis = " & sphr.Axis
  minoraxis = CDbl(sphr.Axis) * (1 - CDbl(sphr.flattening))
  Label3.Caption = "MinorAxis = " & Format(minoraxis, "#########0")
End Sub
```

If you try this, you will see that the values for the major and minor axes of NAD27 and NAD83 are as shown in table 17-1.

Projection and Geographic Coordinate Systems

There are many projections. Some preserve area, some direction, and some distance form one or two points. However, all projections will distort at least one, perhaps all, of these attributes. In this section, you will work with another MO2 object, the *ProjCoordSys* object, which has a *GeoCoordSys* object as one of its members. If you go to the MO2 Help file and look up projections, there is a simplified diagram of the relationship between projection coordinate systems and geographic coordinate systems (see figure 17-4).

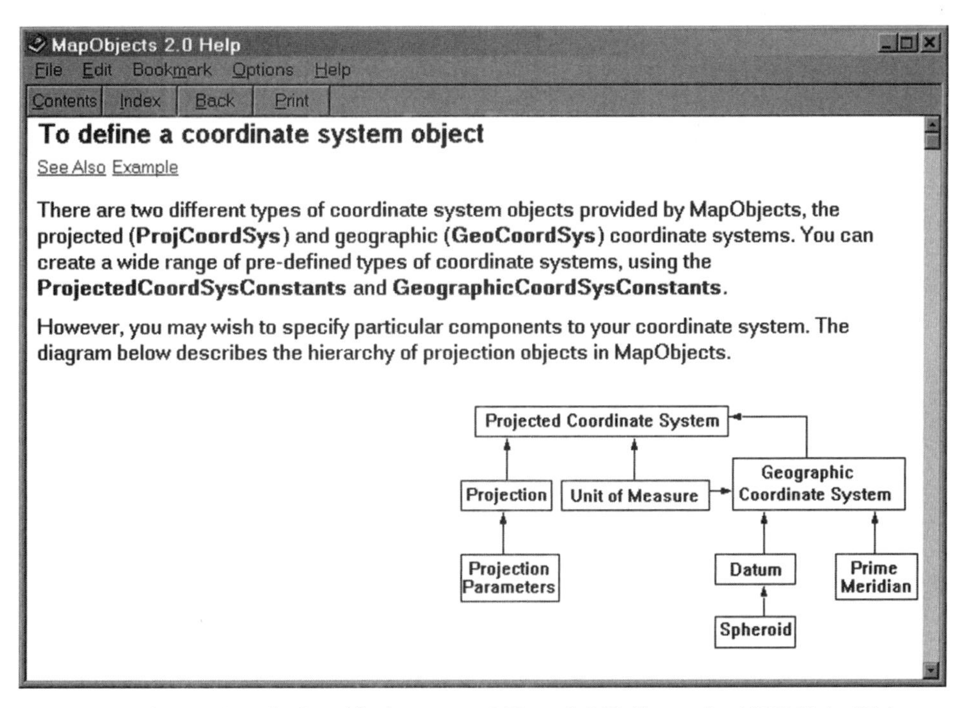

Fig. 17-4. Relationship between PCS and GCS (from the MO2 Help file).

Note that for a projected coordinate system you must have a projection (e.g., Lambert), a unit of measure (such as meters), and a GCS. Each of these input objects, in turn, has its own inputs. For example, you saw in the previous section that a GCS had a *datum* object as a member, and that the *datum* object had a *spheroid* object as a member. You also learned that a spheroid has axis and flattening properties. In the section that follows you will use these objects to assemble shapes that have coordinate systems.

Using Coordinate Systems

Now that you know a little about coordinate systems, both projected and unprojected, you can begin to use them in your existing project. In Chapter 16, you used two shapes, *knoxtract.shp* and *recycle.shp*. Both of these are unprojected; that is, they are in latitude and longitude coordinates based on NAD83. The datum was known because these shapes were derived from TIGER 97 files.

Another shape, *knoxproj.shp*, contains the same information as *knoxtract.shp*, except that it is in Tennessee State Plane Coordinates for 1983. Suppose you wanted to use two of these shapes (one projected and one unprojected) in the same map, or to present unprojected shapes in a projected map window. What things would you have to consider?

The first thing to realize is that you must know the coordinate system of each shape and of the map canvas. That is, like shapes, map controls can have coordinate systems. The key is that MO2 will take such information and display the shapes in the map in a consistent manner.

Let's begin by altering the program you developed in Chapter 16 to handle this situation. Assume that if a shape does not have a corresponding projection file (**.prj* file), it is in geographic coordinates based on NAD83. If you work in another part of the world, this assumption may not be reasonable. Assume also that the map is to display features in the Tennessee State Plane 1983 projection.

Again, when working somewhere else, this assumption might not be reasonable. In fact, you must be very careful when choosing a map projection. The wrong projection can result in parts of your map "disappearing" (areas being projected into a point), or perhaps in errors (divide-by-zero errors). It is a fairly easy task to make someone's mapping program crash or to generate strange-looking results simply by changing the projection.

CD-ROM NOTE: *The VB project in the* Chapter17_2 *directory on the companion CD-ROM starts here.*

1 Load the project you created in Chapter 16. Edit the *Form_Load* sub in form *Form1* to read as follows.

```
Private Sub Form_Load()
    bolFirstLoad = True
    RefreshCombo1
End Sub
```

2 Add the following declaration in *modUtility.bas.*

```
Public bolFirstLoad As Boolean
```

3 Go to form *Form2*'s code page. Find the *addShapeFile* sub.

From now on, when you add a shape, you will check if it is projected. If it is not, assume that it uses the NAD83 datum. That is, assume it is a *moGeoCS_NAD1983* coordinate system. If you did not want to make such an assumption, you could give the user a combo box, similar to *cboGCS* in the previous project, and let them choose the proper coordinate system.

If the shape is projected—that is, if it has a *prj* file—you will use the projection as described in the *prj* file. If you wanted to allow for shapes that were projected but did not have a *prj* file, you could create a combo box of possible projections and let the user select one. You might want to look at the MO2 sample program *Projection* for an example.

Finally, once you have a shape in the map control, you can set the map's coordinate system. That is, each layer will have a coordinate system and the map control will have a coordinate system. Making all these systems work together is MO2's job. You simply have to set things up properly. The following is the new *addShapeFile* sub, with additions indicated in bold.

4 Make the following changes indicated in bold.

```
Private Sub addShapeFile(basepath As String, shpfile As String)
 'This procedure validates and adds a shape file to the
 'Layers collection.
 Dim dCon As New DataConnection
 Dim gSet As GeoDataset
 dCon.Database = basepath
'Set Database property of DataConnection
```

```
 If dCon.Connect Then
  shpfile = GetFirstToken(shpfile, ".")  'Extract suffix of shpfile
  'string
  Set gSet = dCon.FindGeoDataset(shpfile)
   'Find shapefile as GeoDataset in DataConnection
  If gSet Is Nothing Then
   MsgBox "Error opening shapefile " & shpfile
   Exit Sub
  Else
   Dim newLayer As New MapLayer
   newLayer.GeoDataset = gSet
'Set GeoDataset property of new MapLayer
   newLayer.Name = shpfile
'Set Name property of new MapLayer
   Form1.Map1.Layers.Add newLayer
'Add MapLayer to Layers collection
   newLayer.Tag = basepath & "\" & shpfile & ".shp" & "|" & ""
  End If
  'Check for projection information for this layer
  If newLayer.CoordinateSystem Is Nothing Then
    'assume it is lat-long in NAD 83
    Dim LayerCoordSys As New MapObjects2.GeoCoordSys
    Set LayerCoordSys = Nothing
    LayerCoordSys.Type = moGeoCS_NAD1983
    'Set the coordsys for the Layer
    newLayer.CoordinateSystem = LayerCoordSys
  End If
 Else
  MsgBox ConnectErrorMsg(dCon.ConnectError), vbCritical, _
      "Connection error"
 End If
 If bolFirstLoad Then
  Dim MapProjSystem As New MapObjects2.ProjCoordSys
  Set MapProjSystem = Nothing
  MapProjSystem.Type = moProjCS_NAD1983SPCS_TN
  Form1.Map1.CoordinateSystem = MapProjSystem
  bolFirstLoad = False
 End If
End Sub
```

The new code checks to see if the shape file has a *prj* file (i.e., *CoordinateSystem = Nothing?*). If it is *Nothing*, the program creates an empty GCS and sets its type to *moGeoCS_NAD1983*. You should note that as this line is typed in, VB will prompt you with a list of all possible values of geographic coordinate systems (see figure 17-5). Note that you do not need an *Else* statement for the *If* on checking for a coordinate system. This is because the only way you could get an *Else* condition would be if there were a *prj* file. In this case, there is no need to set the coordinate system because you already know what it is!

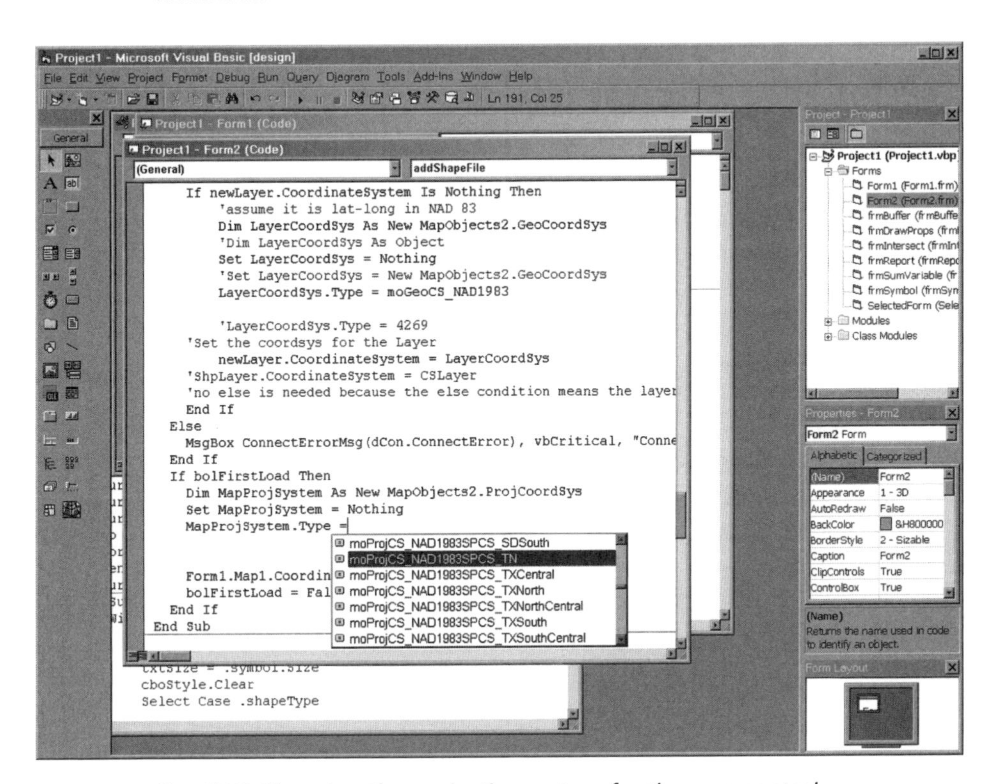

Fig. 17-5. Choosing the projection system for the map control.

Finally, the program checks to see if this is the first layer loaded into the map. *bolFirstLoad* is a Boolean declared as Public at the top of *modUtility*'s code page. In form *Form1*'s *Form_Load* function, the value of *bolFirstLoad* is set to False. Thus, after the first time you load a shape, *bolFirstLoad* will be True. The program creates a *ProjCoordSys* and sets its type to *moProjCS_NAD1983SPPCS_TN* (Tennessee State Plane for 1983). The program then sets the map's coordinate system to this system and the Boolean to False. Again, VB will prompt you with the possible values for the projection type.

It is now possible to mix projected and unprojected shapes in the same map. MO2 will take care of the conversion problems for you. For example, if you load *Knoxtract.shp* and *Knoxproj.shp* in the map, they will display as coincident features, even though the first shape is in latitude and longitude, the second is in feet, and the map control is in meters!

If you use the measure option (click on the map without clicking on a tool), you can draw a line and get its length back in map units (meters). However, if you try the overlay tool developed in Chapter 16, something unexpected will happen. For example, if you overlay the buffers of the *recycle* shape on the *knoxtracts* shape, the buffers will be generated and populations will be estimated, but no buffers will be drawn. If instead of using *knoxtracts* as the polygon shape to overlay you use the *knoxproj* shape, not only will the buffer shapes not be drawn but all population estimates will be zero! Can you guess why? The answers, and fix, await in Chapter 18.

Summary

The ability to project on the fly is a new feature in MapObjects 2. With successive releases of MO2, such as MO2a and MO2.1, ESRI has made the loading and projecting algorithms more efficient. Efficiency aside, the proper use of map projections and coordinate systems requires that you understand how datums, spheroids, and projections are related.

In the first VB project in this chapter you saw how, for any given datum, you can manipulate the parameters for its corresponding spheroid to determine the semi-major and semi-minor axes. In the second project, you were able to combine shapes with differing projections and coordinate systems and properly display them in a map control that used different coordinate units. With these new abilities, you can now fully implement the buffer and overlay functions developed in Chapter 16 so that they will work with any shape.

Chapter 18

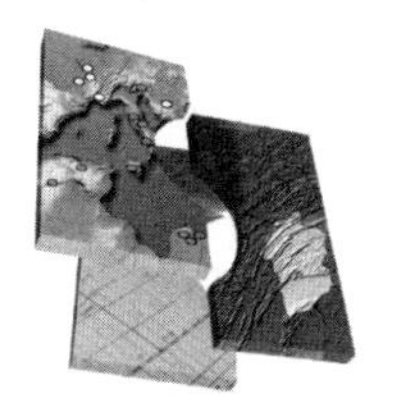

Buffering and Overlay: Part 2

Introduction

At the end of the previous chapter, it was stated that the project for Chapter 17 would not draw the buffers around points. Further, if the polygon layer you overlaid with buffers were *knoxproj*, the population estimates would all be zero. If the polygon layer were *knoxtract* (the unprojected latitude/longitude system based on NAD83), the population estimates could be made. However, hidden in that estimate is the fact that the estimate would be based on area calculations that assume latitude/longitude is a rectangular coordinate system, even though the map display is in projected (state plane) coordinates.

The key concept here is that methods that operate on a shape, such as *SearchShape* or *Intersect*, by default, work on the shape *as it is stored on the disk*—not as it is displayed on the map. Thus, you have to make sure your calculations are using a consistent coordinate system. This is the task you will study in this chapter. Let's look at the code you used to create a point buffer.

```
Dim ellipseRect As New MapObjects2.Rectangle
Set ellipseRect = Nothing
Set pt = recs.Fields("Shape").Value
yoffset = degpermile * buffsize
radians = Abs(pt.Y * pi / 180#)
xoffset = degpermile * buffsize / Cos(radians)
ellipseRect.Bottom = pt.Y - yoffset
ellipseRect.Top = pt.Y + yoffset
ellipseRect.Left = pt.X - xoffset
ellipseRect.Right = pt.X + xoffset
```

There are two things that should catch your attention. The first is that you are calculating a bounding rectangle for each ellipse. Why did you do this? Because you had assumed (in Chapter 16) that the coordinates were in latitude/longitude and that you were correcting for the changing length covered by one degree of longitude as one moved away from the equator.

However, if the map is projected into a rectangular coordinate system, you should be able to use much simpler calculations. In fact, MapObjects 2 comes with a new method, *Buffer.* This method will work fine on several types of features (for example, points, lines, and polygons) as long as the coordinate system used is rectangular. The second thing to note is the line

```
Set pt = recs.Fields("Shape").Value
```

This gets the coordinates of a shape corresponding to the current record. The key here is that *the coordinates are in the units in which they are stored in the shape, not the map control's units.* In the following you will use a method of the *GeoCoordSys* and *ProjCoordSys* objects, called *Transform.* This will transform a shape to the coordinate system you need.

Allowing Buffers of Any Shape Type

CD-ROM NOTE: *The VB project in the* Chapter18_1 *directory on the companion CD-ROM starts here.*

The new *BufferPoints* sub will use the two new methods discussed previously: *Transform* and *Buffer.* As you will see, buffering becomes much easier. In fact, it is so easy that you will now allow the user to buffer point, line, or polygon layers. To do this, you must first change the rules for enabling the buffer button on the toolbar.

1 Change the *RefreshCombo1* sub to read as follows.

```
Public Sub RefreshCombo1()
 Dim i As Integer
 Dim curselected As String
 If (Combo1.ListIndex >= 0) Then
  curselected = Combo1.List(Combo1.ListIndex)
  Toolbar1.Buttons("Identify").Enabled = True
  Toolbar1.Buttons("Select").Enabled = True
  cmdSave.Enabled = True
```

```
  If (Map1.Layers.Count > 1) Then
   cmdIntersect.Enabled = True
   Toolbar1.Buttons("PointBuffer").Enabled = True
  Else
   cmdIntersect.Enabled = False
   Toolbar1.Buttons("PointBuffer").Enabled = False
  End If
 Else
  curselected = ""
  Toolbar1.Buttons("Identify").Enabled = False
  Toolbar1.Buttons("Select").Enabled = False
  Toolbar1.Buttons("PointBuffer").Enabled = False
  cmdSave.Enabled = False
  cmdIntersect.Enabled = False
 End If
 Combo1.Clear
 If Map1.Layers.Count = 0 Then
  Exit Sub
 End If
 For i = 0 To Map1.Layers.Count - 1
  Combo1.AddItem Map1.Layers(i).Name
  If Combo1.List(i) = curselected Then
   Combo1.ListIndex = i
  End If
 Next i
End Sub
```

Here, the program no longer checks to see if the active layer is a point layer. The Buffer button is enabled whenever the number of layers is greater than 1 and there is an active layer. One thing to note is that this code leaves the name of the tool as *PointBuffer* and the routine it calls as *BufferPoints*. If you want to be more disciplined, you should change the names of these to reflect the fact that any feature is now being buffered. You also need to change the *Combo1_Click* sub to reflect the removal of the buffering of only points.

2 Edit the *Combo1_Click* sub to read as follows.

```
Private Sub Combo1_Click()
 Dim i As Integer
 If Combo1.ListIndex >= 0 Then
  Set ActiveLayer = Map1.Layers(Combo1.ListIndex)
  Toolbar1.Buttons("Identify").Enabled = True
  Toolbar1.Buttons("Select").Enabled = True
  cmdSave.Enabled = True
  If (Map1.Layers.Count > 1) Then
```

```
      cmdIntersect.Enabled = True
      Toolbar1.Buttons("PointBuffer").Enabled = True
     End If
    End If
   End Sub
```

You have now set the environment for allowing buffering of any type of layer. However, there is one more thing to consider. Suppose the overlay layer has no numeric fields? Suppose you only want to count the number of features in each buffer? To allow for this, you need to make the following changes in the *frmSumVariable*'s *Form_Load* and *cmdFinish_Click* subs.

3 Make the following changes indicated in bold.

```
Private Sub Form_Load()
 Dim recs As New MapObjects2.Recordset
 Dim i As Integer
 Dim fld As MapObjects2.Field
 Set recs = Form1.OverlayLayer.Records
 cboVariables.AddItem "Count"
 For Each fld In recs.Fields
  If fld.Type = moLong Or fld.Type = moDouble Then
   If fld.Name <> "FeatureId" Then
    cboVariables.AddItem fld.Name
   End If
  End If
 Next
 cboVariables.ListIndex = 0
End Sub
Private Sub cmdFinish_Click()
 Dim recs As New MapObjects2.Recordset
 Dim i As Integer
 Dim fld As MapObjects2.Field
 Set recs = Form1.OverlayLayer.Records
 'We will assume that if no field is chosen from the list of fields
 'then we will simply count the number of features in the buffer
 Set Form1.SumVariable = Nothing
 For Each fld In recs.Fields
  If fld.Name = cboVariables.List(cboVariables.ListIndex) Then
   Set Form1.SumVariable = fld
  End If
 Next
 Form1.bolBufferOverlay = True
 Unload Me
End Sub
```

Buffering

Now that you have set conditions for the Buffer button's behavior and have made the changes to *frmSumVariable*'s code so that a count of items in the buffer is possible, you can turn to the buffering sub itself. MO2's *Buffer* method can be applied to several types of features, including a point, a line, a polygon, a points collection, an ellipse, and a rectangle. You will add the first three cases. However, you cannot simply call the *Buffer* method. You must also make sure that the point, line, or polygon you are buffering has been transformed to the Map control's coordinate system. The following is the code that accomplishes these tasks.

This code begins by declaring the necessary variables. Because the Map control uses Tennessee State Plane 83 coordinates, the map units will be in meters. If you did not know this, you could always get the units from the *ProjCoordSys* object's *Unit* method. This will return one of several possible values. Figure 18-1 shows these values, which are listed in MO2 Help file.

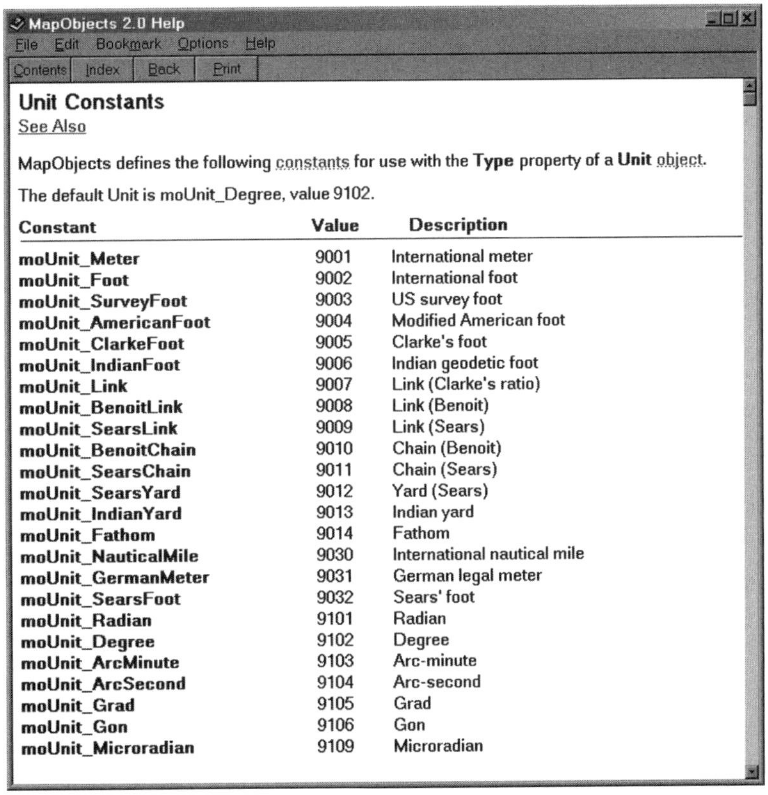

Fig. 18-1. Possible unit values.

MapObjects 2.0 Help
File Edit Bookmark Options Help
Contents Index Back Print

Unit Constants
See Also

MapObjects defines the following constants for use with the **Type** property of a **Unit** object.

The default Unit is moUnit_Degree, value 9102.

Constant	Value	Description
moUnit_Meter	9001	International meter
moUnit_Foot	9002	International foot
moUnit_SurveyFoot	9003	US survey foot
moUnit_AmericanFoot	9004	Modified American foot
moUnit_ClarkeFoot	9005	Clarke's foot
moUnit_IndianFoot	9006	Indian geodetic foot
moUnit_Link	9007	Link (Clarke's ratio)
moUnit_BenoitLink	9008	Link (Benoit)
moUnit_SearsLink	9009	Link (Sears)
moUnit_BenoitChain	9010	Chain (Benoit)
moUnit_SearsChain	9011	Chain (Sears)
moUnit_SearsYard	9012	Yard (Sears)
moUnit_IndianYard	9013	Indian yard
moUnit_Fathom	9014	Fathom
moUnit_NauticalMile	9030	International nautical mile
moUnit_GermanMeter	9031	German legal meter
moUnit_SearsFoot	9032	Sears' foot
moUnit_Radian	9101	Radian
moUnit_Degree	9102	Degree
moUnit_ArcMinute	9103	Arc-minute
moUnit_ArcSecond	9104	Arc-second
moUnit_Grad	9105	Grad
moUnit_Gon	9106	Gon
moUnit_Microradian	9109	Microradian

Once you know the units, you can convert the buffer distance to the appropriate value.

1 Edit the *BufferPoints* sub to read as follows.

```
Public Sub BufferPoints()
 Dim buffsize As Double
 Dim recs As New MapObjects2.Recordset
 Dim prjpt As New MapObjects2.Point
 Dim prjline As New MapObjects2.line
 Dim prjpoly As New MapObjects2.Polygon
 Dim i As Integer
 Dim meterpermile As Double
 meterpermile = 1609.344
 buffsize = CDbl(BufferDistance) * meterpermile
```

The program then needs to get the buffer layer's *recordset* and clear out the *colEllipseRect* collection. As with the name *BufferPoints*, the *colEllipseRect* collection is not aptly named. You will be storing polygons of many possible shapes. Still, a collection by any other name will still store polygons.

2 Continuing with *BufferPoints*, enter the following.

```
 Set recs = BufferLayer.Records
 recs.MoveFirst
 If colEllipseRect.Count > 0 Then
  For i = 1 To colEllipseRect.Count
   colEllipseRect.Remove (1)
  Next
 End If
 Dim CS As Object
 Set CS = Map1.CoordinateSystem
```

The next section of code cycles through each record in the record set. It creates a new polygon to hold the buffer. Then, depending on the type of feature being buffered, it gets its location coordinates. These are then transformed to the Map control's coordinate system from the buffer layer's coordinate system. Once the buffer is created, the program adds it to the collection and cycles through the records. When this process is finished, the sub is ended.

3 Enter the following below the lines you entered in step 2.

```
Do While Not recs.EOF
 Dim buffpoly As New MapObjects2.Polygon
 Select Case BufferLayer.shapeType
 Case moPoint
```

```
    Set prjpt = CS.Transform(BufferLayer.CoordinateSystem, _
        recs.Fields("Shape").Value)
    Set buffpoly = prjpt.Buffer(buffsize, Map1.FullExtent)
   Case moLine
    Set prjline = CS.Transform(BufferLayer.CoordinateSystem, _
        recs.Fields("Shape").Value)
    Set buffpoly = prjline.Buffer(buffsize, Map1.FullExtent)
   Case moPolygon
    Set prjpoly = CS.Transform(BufferLayer.CoordinateSystem, _
        recs.Fields("Shape").Value)
    Set buffpoly = prjpoly.Buffer(buffsize, Map1.FullExtent)
   End Select
   colEllipseRect.Add buffpoly
   recs.MoveNext
  Loop
End Sub
```

There are two new methods used in this sub, *Transform* and *Buffer*. *Transform* will take a coordinate system (in this case, *BufferLayer*'s coordinate system) and an object (the shape corresponding to the current record in the record set) and transform that shape to the object's (*CS=Map1.CoordinateSystem*) coordinate system.

Next, the transformed point, line, or polygon is buffered to the *buffsize*. Note the second argument in each buffer call. MO2 requires an extent sent that exceeds the largest extent of the buffers. The MO2 Help file uses the Map control's fullest extent. However, this is not always enough. Consider figure 18-2. In this figure, a line shape—the *interstates* shape—has been buffered. However, not all lines have buffers. This is because these buffers exceed the extent listed in the call to *Buffer* (*Map1.FullExtent*). You need to correct the previous case.

4 Insert the following lines before the start of the *Do While* loop.

```
Dim rect As New MapObjects2.Rectangle
Set rect = Map1.FullExtent
rect.ScaleRectangle (2)
```

5 In each buffer call, replace *Map1.FullExtent* with *rect*. Thus, the loop should now read as follows.

```
Do While Not recs.EOF
  Dim buffpoly As New MapObjects2.Polygon
  Select Case BufferLayer.shapeType
  Case moPoint
   Set prjpt = CS.Transform(BufferLayer.CoordinateSystem, _
     recs.Fields("Shape").Value)
```

```
   Set buffpoly = prjpt.Buffer(buffsize, rect)
  Case moLine
   Set prjline = CS.Transform(BufferLayer.CoordinateSystem, _
       recs.Fields("Shape").Value)
   Set buffpoly = prjline.Buffer(buffsize, rect)
  Case moPolygon
   Set prjpoly = CS.Transform(BufferLayer.CoordinateSystem, _
       recs.Fields("Shape").Value)
   Set buffpoly = prjpoly.Buffer(buffsize, rect)
  End Select
  colEllipseRect.Add buffpoly
  recs.MoveNext
Loop
```

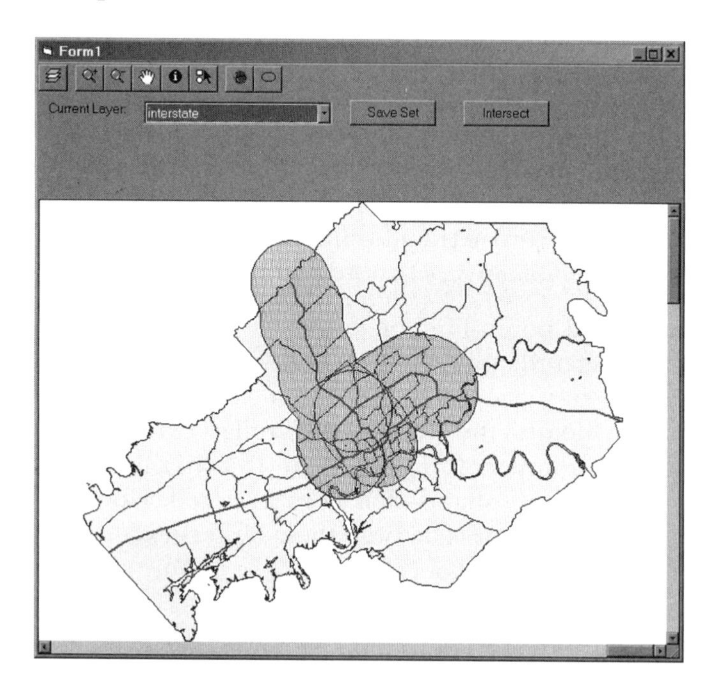

There are two things to notice in this code. First, because the rectangle is based on the Map control's coordinate system, you do not have to transform it. It is in the correct units. Second, as a programmer, you must know the maximum size of rectangle you will need to hold the buffers your users will create. If it is not large enough, you will end up with a case similar to that shown in figure 18-2.

Fig. 18-2. Buffered interstates shape.

Reporting the Results

There is one last task in completing the buffer and overlay function. You need to calculate the intersection of the overlay layer and the buffers. You do this in the *AfterLayerDraw* sub. The first part of this sub remains unchanged. You only need to focus on the section that starts with

```
If index = 0 And bolBufferOverlay Then
```

If these conditions are met, the buffers should be drawn. You also need to calculate the intersection of the buffer and the overlay layer. However, you must be aware of some new problems. First, all spatial calculations must use the same projected coordinate system. Second, you must account for the option of choosing *Count* as the summary variable. Finally, you can no longer assume the features being buffered are the stores with recycle centers.

The program starts by declaring the objects and variables needed within this *If* statement. Note that it declares a rectangle similar to that used in the buffer sub. Like the one in that sub, this rectangle must be large enough to hold all features to be drawn.

1 Enter the following.

```
Dim poly1, poly2 As New MapObjects2.Polygon
Dim rect As New MapObjects2.Rectangle
Set rect = Map1.FullExtent
rect.ScaleRectangle (2)
Dim buffpoly As New MapObjects2.Polygon
Dim buffersym As New MapObjects2.symbol
Dim reptext As String
```

The program then needs to set the buffer symbol and initialize the report string. You can no longer assume that there is a summation variable. It needs to be checked. If *SumVariable* is nothing, you know that the summation is a simple count of all features in the overlay layer that lie within the buffers.

2 Enter the following.

```
buffersym.Color = moRed
buffersym.Style = moLightGrayFill
If Not SumVariable Is Nothing Then
 reptext = "The summation variable is " & SumVariable.Name & vbCrLf
Else
 reptext = "The summation variable is count" & vbCrLf
End If
reptext = reptext + "The buffer distance is " & BufferDistance & _
    " miles" & vbCrLf & vbCrLf
```

The next step is to get the records in the buffer layer and the Map control's coordinate system.

3 Enter the following.

```
Dim buffrecs As MapObjects2.Recordset
Set buffrecs = BufferLayer.Records
buffrecs.MoveFirst
```

```
'We need the CS for the transforms
Dim CS As Object
Set CS = Map1.CoordinateSystem
```

You can now operate on each buffer in the collection. Note that new code is needed to handle the *SumVariable Is Nothing* case. Because you are no longer assuming that the recycle centers are the buffered layer, you cannot use *store name* as a way of identifying each feature. Instead, you will use each feature's *featureId.* You must also project the overlay layer's features into the Map control's coordinate system.

You must do this because the buffer polygons were generated in the Map control's coordinate system. This means that you must use a consistent coordinate system when calculating the intersection of the two features (the current buffer polygon and the shape for the current record in the overlay layer). Finally, because the overlay layer is no longer assumed to be a polygon, you must adjust your calculations of the *SumVariable*'s value based on whether the overlay layer is a point, line, or polygon layer.

4 Continue editing the *AfterLayerDraw* sub to read as follows.

```
For i = 1 To colEllipseRect.Count
 Set buffpoly = colEllipseRect.Item(i)
 'Uncomment the next line if you want to see each buffer
 Map1.DrawShape buffpoly, buffersym
 If i = 1 Then
  Set poly2 = buffpoly
 Else
  Set poly2 = poly1.Union(buffpoly, rect)
 End If
 Set poly1 = poly2
 Dim curfeature, curfeatureprj As Object
 Dim selfeature As Object
 Dim allrecs As MapObjects2.Recordset
 Set allrecs = OverlayLayer.Records
  allrecs.MoveFirst
  Dim popest As Double
  popest = 0
  Do While Not allrecs.EOF
   Set curfeature = allrecs("Shape").Value
   'As with the buffer points we must ensure that the tracts
   'polygons are transformed to the proper coordinate system and we
   'must handle points, lines, and polygons as special cases
   Set curfeatureprj = CS.Transform(OverlayLayer.CoordinateSystem,_
      curfeature)
```

```
   Set selfeature = curfeatureprj.Intersect(buffpoly)
   If Not selfeature Is Nothing Then
     If Not SumVariable Is Nothing Then
       Select Case OverlayLayer.shapeType
       Case moPoint
         popest = popest + allrecs(SumVariable.Name)
       Case moLine
         popest = popest + allrecs(SumVariable.Name) * _
            selfeature.Length / curfeatureprj.Length
       Case moPolygon
         popest = popest + allrecs(SumVariable.Name) * _
            selfeature.Area / curfeatureprj.Area
       End Select
     Else
       popest = popest + 1
   End If
  End If
  allrecs.MoveNext
 Loop
 'Note the change from store name to feature id in the next line
 reptext = reptext + "The estimated value within the buffer of " & _
   buffrecs.Fields("featureId").Value & " is " & Format(popest, _
   "######0.0") & vbCrLf
 buffrecs.MoveNext
Next
```

At this point you have processed all buffers. All you need to do now is process the union of all buffers.

5 Enter the following.

```
Map1.DrawShape poly2, buffersym
popest = 0
allrecs.MoveFirst
Do While Not allrecs.EOF
 Set curfeature = allrecs("Shape").Value
 Set curfeatureprj = CS.Transform(OverlayLayer.CoordinateSystem, _
     curfeature)
 Set selfeature = curfeatureprj.Intersect(poly2)
 If Not selfeature Is Nothing Then
  If Not SumVariable Is Nothing Then
   Select Case OverlayLayer.shapeType
   Case moPoint
    popest = popest + allrecs(SumVariable.Name)
   Case moLine
    popest = popest + allrecs(SumVariable.Name) * selfeature.Length _
      / curfeatureprj.Length
```

```
   Case moPolygon
    popest = popest + allrecs(SumVariable.Name) * selfeature.Area / _
      curfeatureprj.Area
   End Select
  Else
   popest = popest + 1
  End If
 End If
 allrecs.MoveNext
Loop
reptext = reptext + vbCrLf + "The estimated value within all _
   buffers is " & Format(popest, "######0.0")
frmReport.txtReport = reptext
End If
End Sub
```

You can now buffer points, lines, or polygons, and overlay these buffers on point, line, or polygon layers. Try loading the *knoxprj*, *interstate*, and *recycle* layers into your program. Set each one as the active layer, buffer it, and use one of the other layers as the overlay layer. You should be able to generate results like those shown in figure 18-3.

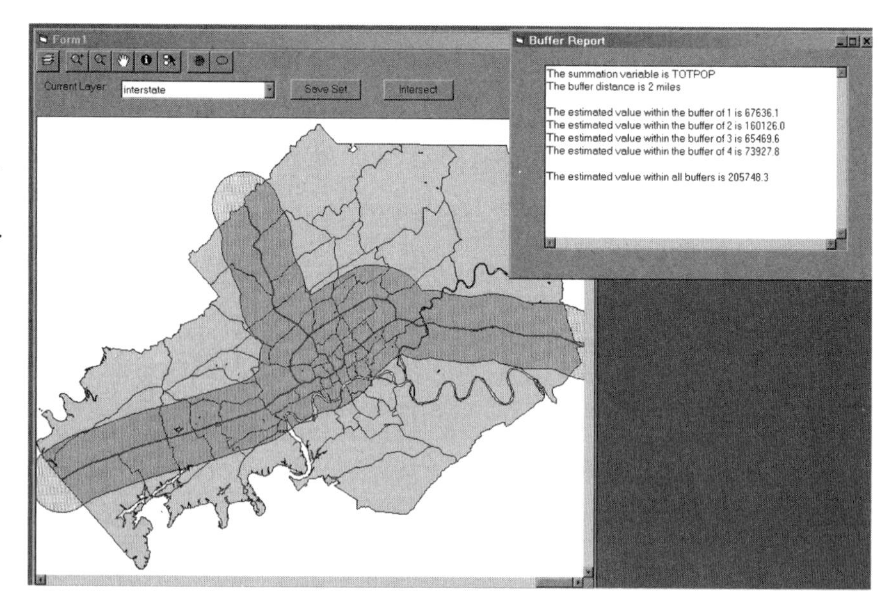

Fig. 18-3. Buffering the interstate *layer and estimating the population.*

Tying Up Some Loose Ends

There are a few problems you might have noticed with the current project. If you try to use the Identify or Select tools, they probably will not work. This is because they were written before you used differing coordinate systems. For example, the Identify tool (coded in the *MouseDown* event) finds features in a rectangle using the following approach.

```
Set r = Map1.TrackRectangle
Set recs = ActiveLayer.SearchShape(r, moAreaIntersect, "")
```

Here, there is no guarantee that the active layer and the Map control will have the same coordinate system. A further complication is that in Chapter 14 you changed the scale mode of the project from twips to pixels. You did this because it made working with the click location on an image easier to capture and manipulate. However, when you switch to pixels from twips, the *ToMapPoint* function does not behave as you might expect. Fortunately, there is an easy way to handle this case.

Let's begin by considering the changes you need to make to the approach represented by the previous code. The rectangle *r* will be in the Map control's units, even if these are projected units. For the current project, this is Tennessee State Plane 1983. The active layer will have its own coordinate system, which may not match that of the Map control. Therefore, you need to transform the resulting rectangle from the Map control's coordinate system to the active layer's coordinate system. If you look at the buffering sub, you will see that this is the opposite of what you did in that case. There, you transformed the buffer layer's shapes into the Map control's coordinate system.

To make the necessary changes, you need to declare a new object to hold the active layer's coordinate system. Thus, you need to modify the top of this sub as follows.

1 Make the following changes indicated in bold.

```
Private Sub Map1_MouseDown(Button As Integer, Shift As Integer, X _
    As Single, Y As Single)
 Dim curRectangle As New MapObjects2.Rectangle
 Dim i As Integer
 Dim curIndex As Integer
 Dim allrecs As MapObjects2.Recordset
 Set allrecs = ActiveLayer.Records
```

```
Dim r As MapObjects2.Rectangle
Dim rprj As MapObjects2.Rectangle
Dim pt As New MapObjects2.Point
Dim ptprj As New MapObjects2.Point
Dim CS As Object
Set CS = ActiveLayer.CoordinateSystem
```

You will use *rprj* to hold the rectangle transformed from the Map control's coordinate system. Thus, the section of the sub that deals with tracking the rectangle for the identify tool should read as follows.

2 Make the following changes indicated in bold.

```
ElseIf Toolbar1.Buttons("Identify").Value = 1 Then
  Dim recs As MapObjects2.Recordset
  Dim tol As Double
  Set r = Map1.TrackRectangle
  Set rprj = CS.Transform(Map1.CoordinateSystem, r)
  Set recs = ActiveLayer.SearchShape(rprj, moAreaIntersect, "")
```

Note that this code transforms the rectangle from the Map control's coordinate system to the active layer's coordinate system. It then uses the transformed rectangle in the *SearchShape* method.

The *Identify* method also contains code for handling cases in which the user clicks on the map rather than dragging a rectangle. The following code *may* work.

```
  If recs.EOF Then
   tol = 250 'set tolerance in meters
   Dim rect As New MapObjects2.Rectangle
   Set rect = Map1.FullExtent
   rect.ScaleRectangle (2)
   Set pt = Map1.ToMapPoint(X, Y)
   Dim ptbuffer As New MapObjects2.Polygon
   Dim prjbuffer As New MapObjects2.Polygon
   Set ptbuffer = pt.Buffer(tol, rect)
   Set prjbuffer = CS.Transform(Map1.CoordinateSystem, prjbuffer)
   Set recs = ActiveLayer.SearchShape(prjbuffer, moAreaIntersect, "")
  End If
```

Note that this code does not set the tolerance in screen units, but in the units of the Map control's coordinate system. This works because the Map control has a coordinate system set in meters. Thus, there is no call to *ToMapDistance*. This code buffers the point of the mouse click, transforms the buffer into the active layer's coordinate system, and then uses the *SearchShape* command to get the resultant record set.

If form *Form1*'s *ScaleMode* is in twips, the previous code works fine. However, you changed form *Form1*'s mode to pixels. In this case, *ToMapPoint* will not return the proper point. This is because pixels are not square, nor are they of a uniform size. Fortunately, there is an easy fix. From the first part of the code, you know that *Track-Rectangle* returns a shape in the Map control's coordinate system. You can simply treat a mouse click as a degenerate rectangle. The following is the new code.

3 Enter the following.

```
tol = 250 'set tolerance in meters
   Dim rect As New MapObjects2.Rectangle
   Set rect = Map1.FullExtent
   rect.ScaleRectangle (2)
   'Set pt = Map1.ToMapPoint(X, Y)
   'Instead of using ToMapPoint, we will get the center of
   'the "rectangle"
   pt.X = r.Center.X
   pt.Y = r.Center.Y
   Dim ptbuffer As New MapObjects2.Polygon
   Dim prjbuffer As New MapObjects2.Polygon
   Set ptbuffer = pt.Buffer(tol, rect)
   Set prjbuffer = CS.Transform(Map1.CoordinateSystem, ptbuffer)
   Set recs = ActiveLayer.SearchShape(prjbuffer, moAreaIntersect, "")
```

This version of the code will work no matter what form *Form1*'s scale mode is in.

There is one final loose end to take care of: the *Spatial_Select* sub needs to be modified so that it will work with these projected shapes. In previous chapters (starting from the project in the *Chapter 12_3* directory on the companion CD-ROM), you used the *SearchShape* method to find all elements in the active layer that intersect with selected records of the *intersect* layer. This was accomplished with the following code (from the *Chapter 12_3* project).

```
Public Sub Spatial_Select(lyrname As String)
  Dim curIndex, refIndex As Integer
  Dim arec As MapObjects.Recordset
  Dim i As Integer
  curIndex = -1
  refIndex = -1
  If (colRecSetClass.Count = 0) Then
   Dim bLayer As New clsRecSet
   bLayer.Name = ActiveLayer.Name
   colRecSetClass.Add bLayer
```

```
  curIndex = 1
 Else
  For i = 1 To colRecSetClass.Count
   If colRecSetClass.Item(i).Name = ActiveLayer.Name Then
    curIndex = i
   End If
   If colRecSetClass.Item(i).Name = lyrname Then
    refIndex = i
   End If
   If (curIndex >= 0) And (refIndex >= 0) Then
    Exit For
   End If
  Next i
 End If
 If (curIndex = -1) Then
  bLayer.Name = ActiveLayer.Name
  colRecSetClass.Add bLayer
  curIndex = colRecSetClass.Count
 End If
 If refIndex = -1 Then
  Dim allrecs As New MapObjects.Recordset
  Dim lyr As MapObjects.MapLayer
  For Each lyr In Map1.Layers
   If lyr.Name = lyrname Then
    Set allrecs = lyr.Records
    Exit For
   End If
  Next
  allrecs.MoveFirst
  Set gSelection = ActiveLayer.SearchShape(allrecs, _
        moAreaIntersect, "")
  gSelection.MoveFirst
  Do While Not gSelection.EOF
   Set arec = ActiveLayer.SearchExpression("featureId = " & _
        gSelection.Fields("FeatureID").ValueAsString)
   colRecSetClass.Item(curIndex).RecSetCol.Add arec
   gSelection.MoveNext
  Loop
 Else
  Dim currec As New MapObjects.Recordset
  For i = 1 To colRecSetClass.Item(refIndex).RecSetCol.Count
   Set currec = colRecSetClass.Item(refIndex).RecSetCol.Item(i)
   Set gSelection = ActiveLayer.SearchShape(currec, _
        moAreaIntersect, "")
   If gSelection.Count > 0 Then
    gSelection.MoveFirst
```

```
     Do While Not gSelection.EOF
      Set arec = ActiveLayer.SearchExpression("featureId = " & _
           gSelection.Fields("FeatureID").ValueAsString)
      colRecSetClass.Item(curIndex).RecSetCol.Add arec
      gSelection.MoveNext
     Loop
    End If
   Next
  End If
  Map1.Refresh
End Sub
```

Here, the *SearchShape* method searches an entire layer to see if a shape or shapes intersect the layer. However, *SearchShape* assumes that the two layers, in this case the active layer and the intersect layer, are stored in the same coordinate system. You can no longer assume that this is the case.

To overcome this problem, you need to transform each shape in the intersect layer to the same coordinate system as that of the active layer. You can then use the *SearchShape* method to find the records in the active layer that touch or intersect the transformed shape. This code follows, with the new key sections indicated in bold.

4 Make the following changes indicated in bold.

```
Public Sub Spatial_Select(lyrname As String)
 Dim CS As Object
 Dim reflayer As New MapObjects2.MapLayer
 Dim ptshp As New MapObjects2.Point
 Dim lineshp As New MapObjects2.Line
 Dim polyshp As New MapObjects2.Polygon
 Dim typeofshape As Integer
 Dim curIndex, refIndex As Integer
 Dim arec As MapObjects2.Recordset
 Dim i As Integer
 Dim lyr As MapObjects2.MapLayer
 Set CS = ActiveLayer.CoordinateSystem
 For Each lyr In Map1.Layers
  If lyr.Name = lyrname Then
   Set CSreflayer = lyr.CoordinateSystem
   Set reflayer = lyr
   typeofshape = lyr.shapeType
   Exit For
  End If
 Next
 curIndex = -1
```

```
refIndex = -1
If (colRecSetClass.Count = 0) Then
 Dim bLayer As New clsRecSet
 bLayer.Name = ActiveLayer.Name
 colRecSetClass.Add bLayer
 curIndex = 1
Else
 For i = 1 To colRecSetClass.Count
  If colRecSetClass.Item(i).Name = ActiveLayer.Name Then
   curIndex = i
  End If
  If colRecSetClass.Item(i).Name = lyrname Then
   refIndex = i
  End If
  If (curIndex >= 0) And (refIndex >= 0) Then
   Exit For
  End If
 Next i
 End If
 If (curIndex = -1) Then
  bLayer.Name = ActiveLayer.Name
  colRecSetClass.Add bLayer
  curIndex = colRecSetClass.Count
 End If
 If refIndex = -1 Then
  Dim allrecs As New MapObjects2.Recordset
  For Each lyr In Map1.Layers
   If lyr.Name = lyrname Then
    Set allrecs = lyr.Records
   Exit For
  End If
  Next
  allrecs.MoveFirst
  Do While Not allrecs.EOF
   Select Case typeofshape
   Case moPoint
    Set ptshp = CS.Transform(reflayer.CoordinateSystem, _
      allrecs.Fields("Shape").Value)
    Set gSelection = ActiveLayer.SearchShape(ptshp, _
      moEdgeTouchOrAreaIntersect, "")
   Case moLine
    Set lineshp = CS.Transform(reflayer.CoordinateSystem, _
      allrecs.Fields("Shape").Value)
    Set gSelection = ActiveLayer.SearchShape(lineshp, _
      moEdgeTouchOrAreaIntersect, "")
   Case moPolygon
```

```
    Set polyshp = CS.Transform(reflayer.CoordinateSystem, _
      allrecs.Fields("Shape").Value)
    Set gSelection = ActiveLayer.SearchShape(polyshp, _
      moEdgeTouchOrAreaIntersect, "")
  End Select
  gSelection.MoveFirst
  Do While Not gSelection.EOF
  Set arec = ActiveLayer.SearchExpression("featureId = " & _
       gSelection.Fields("FeatureID").ValueAsString)
  colRecSetClass.Item(curIndex).RecSetCol.Add arec
  gSelection.MoveNext
  Loop
  allrecs.MoveNext
 Loop
Else
 Dim currec As New MapObjects2.Recordset
 For i = 1 To colRecSetClass.Item(refIndex).RecSetCol.Count
  Set currec = colRecSetClass.Item(refIndex).RecSetCol.Item(i)
  Select Case typeofshape
  Case moPoint
   Set ptshp = CS.Transform(reflayer.CoordinateSystem, _
     currec.Fields("Shape").Value)
   Set gSelection = ActiveLayer.SearchShape(ptshp, _
     moEdgeTouchOrAreaIntersect, "")
  Case moLine
   Set lineshp = CS.Transform(reflayer.CoordinateSystem, _
     currec.Fields("Shape").Value)
   Set gSelection = ActiveLayer.SearchShape(lineshp, _
     moEdgeTouchOrAreaIntersect, "")
  Case moPolygon
   Set polyshp = CS.Transform(reflayer.CoordinateSystem, _
     currec.Fields("Shape").Value)
   Set gSelection = ActiveLayer.SearchShape(polyshp, _
     moEdgeTouchOrAreaIntersect, "")
   End Select
  If gSelection.Count > 0 Then
  gSelection.MoveFirst
  Do While Not gSelection.EOF
   Set arec = ActiveLayer.SearchExpression("featureId = " & _
      gSelection.Fields("FeatureID").ValueAsString)
   colRecSetClass.Item(curIndex).RecSetCol.Add arec
   gSelection.MoveNext
  Loop
  End If
 Next
End If
```

```
  Map1.Refresh
End Sub
```

The new sections test for an intersection with the active layer for each feature in the intersect layer.

Summary

There are some other enhancements that could be made to this program. First, as noted at the end of Chapter 12, the current *Spatial_Select* sub allows for the same feature to be selected multiple times. It would be wise to check for such cases and remove the duplicates. You should probably also incorporate a Command button for clearing existing buffers, if any exist. You might also want to remove elements of *colRecSetClass* for layers that have no selected features. The current code allows a layer to have an entry in the class even when it does not have any selected records.

More advanced options might include allowing buffers for each layer and using the *overlay* function to select a set, much like the functionality of the Intersect button. Finally, when a data set is saved, the program should check to see if the user wishes to save it in its original coordinate system or in the Map control's coordinate system. That is, the user should be asked if the shape should be saved in projected units. MO2 comes with a new method, called *Export*, that makes such a task fairly painless. (See the MO2 Help file or MO2's VB sample project named *Projector.*)

The methods covered in this and the previous two chapters are powerful, but that power comes at a price. As was noted in Chapter 17, loading all projection and coordinate system definitions (all enumeration cases) can be quite slow. In addition, if shapes are quite complex, on-the-fly projection can cause map refreshing to be slow as well. It is probably not something you would want to do when using VB/MO programs as the basis of web-based mapping.

Index